PRAISE FOR *LEADING SUSTAINABLE INNOVATION*

'In a world where sustainability is no longer optional, *Leading Sustainable Innovation* is a vital resource for forward-thinking leaders. Clear, concise and directly actionable, this book will help shape the future of complex, technical sustainable innovation projects, bridging the gap between theory and practice.'

Lucy Armstrong, CEO, The Alchemists and Chair, Port of Tyne

'This book is a breakthrough. No buzzwords, not a wasted word or paragraph. Jo North's book provides accessible, smart, practical and structured advice. Not just on *what* sustainable innovation is, and importantly *how* to do it, but inspirational and accessible information for leaders and those who aspire to leadership on *why* successful sustainable innovation is critical for them and for the health and future of our planet.'

Dr Joanna Berry, Associate Dean for Engagement, Durham University Business School and Director for Business, Innovation and Entrepreneurship, Durham Energy Institute

'A brilliantly well researched and practical step-by-step guide on how to innovate whilst holding true to your values. Jo North peppers the theory with relatable examples which will inspire leaders from all sectors to address their challenges head-on and with sustainability at the core.'

Heidi Mottram CBE, CEO, Northumbrian Water Group

'This will be an invaluable guide for anyone working in infrastructure, construction or the wider industry who is navigating the challenges of leading sustainable innovation. With invaluable insights, toolkits, checklists and case studies, it provides both the theory and the action.'

Julia Prescot, Co-Founder, Meridiam, Deputy Chair, National Infrastructure Commission and Port of Tyne, Chair, Neuconnect and Fulcrum Infrastructure Group and Honorary Professor, The Bartlett School of Sustainable Construction, UCL

'This is a must-read for anyone passionate about merging innovation with environmental and economic responsibility. Jo North offers practical strategies and inspiring case studies that bring her approaches to life and demonstrate how businesses can thrive while prioritising sustainability. Her expertise and forward-thinking approach make this book an invaluable resource for leaders who want to grow their business and innovate their assets in a sustainable way that delivers societal and environmental benefits.'

Amanda Selvaratnam, Associate Director of Research and Enterprise and Head of Enterprise Services, University of York.

'Leading sustainable technical innovation is a tough and complex business. This wonderful book demystifies, simplifies and inspires.'

Patrick Dunne, Chair, Boardelta, and Chair, Royal Voluntary Service

'A one-stop shop for business leaders in how to become a sustainable business. Transformative and inspiring. Essential for leaders of technical programs driving impactful change in today's challenging business environment.'

Joe Lavan, Commercial Director, Coffey Group

'This book offers practical methodologies, tools and a clear roadmap for business leaders to achieve lasting change in their projects and programmes. Essential reading for technical professionals and leaders alike, it empowers them with the knowledge and confidence to identify, drive and embed sustainable growth and resilience in an ever-evolving global landscape.'

Dr Maxine Mayhew, Chief Operating Officer - Collections and Specialist Services, Biffa

'This book very clearly explains the importance of sustainable innovation and provides a clear roadmap for implementing sustainable innovation projects in complex technical environments. It provides a great insight into all building blocks and key ingredients in creating a sustainable ecosystem and how to manage innovation. Critical for students, academics and researchers who aspire to take a leadership role in sustainable innovation.'

Professor Tony Roskilly, Director, Durham Energy Institute, Director, UK National Clean Maritime Research Hub. Academic Lead, Teesside Industrial Cluster, Chair of Energy Systems, Department of Engineering, University of Durham

'There is a lot written and said on concepts like green business models, whole life cycles, and the circular economy. This book shows managers and practitioners how to implement these and how to understand them in their own industries, through a plethora of practical frameworks, tools for thought and measurement and checklists that build into a road map toward sustainable innovation.'

Professor Jonathan Sapsed, Chair of Innovation and Entrepreneurship, Newcastle University Business School

Leading Sustainable Innovation

A roadmap for technical environments

Jo North

First published in Great Britain and the United States in 2024 by Kogan Page Limited

2nd Floor, 45 Gee Street
London
EC1V 3RS
United Kingdom

8 W 38th Street, Suite 902
New York, NY 10018
USA

www.koganpage.com

Kogan Page books are printed on paper from sustainable forests.

ISBNs
Hardback 978 1 3986 1670 7
Paperback 978 1 3986 1668 4
Ebook 978 1 3986 1669 1

British Library Cataloguing-in-Publication Data
A CIP record for this book is available from the British Library.

Library of Congress Control Number
2024023309

Typeset by Integra Software Services, Pondicherry
Print production managed by Jellyfish
Printed and bound by CPI Group (UK) Ltd, Croydon CR0 4YY

CONTENTS

LIST OF FIGURES

FIGURES

ABOUT THE AUTHOR

Jo North PhD is an expert on strategic sustainable innovation within complex technical industries, with experience spanning sectors such as transport, manufacturing, engineering, infrastructure, energy, utilities, nuclear and technology. She is the founder and CEO of The Big Bang Partnership Ltd, where she guides businesses in driving sustainable growth and capitalizing on the unique challenges and opportunities of sustainable innovation in complex STEM industries.

Jo is also the director of technology and transformation at Port of Tyne, and leads the UK's inaugural Maritime Innovation Hub, working towards decarbonizing and digitalizing ports and supply chains and setting new benchmarks for clean energy, social impact and economic growth. In addition, she is an associate in Business Innovation and Creativity at the University of York.

Jo has worked with organizations such as Sage, Microsoft, Aviva, National Nuclear Laboratory, Transport for London and Transdev. As part of her work, she has facilitated sustainable innovation roadmapping for international companies, worked on long-term environment and sustainability strategies for water companies and collaborated with energy companies on their innovation plans. She has also previously held director roles at Northern Rail and East Coast trains. Jo is based near Newcastle, in the UK.

PREFACE

Fortunately, there has been a shift from talking about sustainability and making promises to achieve future targets to taking steps towards achieving them. The easy, inexpensive quick wins are being achieved, meaning that governments, industries and organizations now need to solve tricky and uncertain challenges to make progress. Sustainable innovation is the process of finding new, planet- and people-friendly solutions to these challenges, of breaking new ground and creating a paradigm shift in how organizations do business.

Since I founded my innovation consultancy, The Big Bang Partnership Ltd, 14 years ago, I have been working with leaders of sustainable innovation across a variety of technical sectors to tackle challenges of this nature within their own specific contexts. These include the water and wastewater, energy, logistics and transportation, infrastructure, construction, communications, engineering and technology sectors. Since 2019, I have also been working to support decarbonization and digitalization in the maritime industry in my role as director of technology and transformation at Port of Tyne and as leader of the UK's national 2050 Maritime Innovation Hub. In addition I have worked with many academic researchers and their teams to help them bring their sustainable innovation research into practice.

There are common themes which unite this diversity of sectors in tackling these tricky and uncertain challenges. Each sector is figuring out how to deal with the shared challenges of achieving true organizational sustainability. They're simultaneously navigating energy transition, digital transformation, water and waste management, circularity, being a good corporate citizen and employer, while continuing to perform day-to-day in a volatile, uncertain, complex and ambiguous world. New organizations can begin as they mean to go on, but organizations with history, an established culture, a supply chain, processes and legacy assets need to make what is potentially the biggest, most important transformation they have ever faced.

It is clear to me that, as the saying goes, what got organizations here won't get them there. This historic shift in the fight against climate change needs leaders and their teams to adopt a new mindset, skills, processes and systems. We can use, though not rely on, our experience and previous approaches; but if we are to succeed, we need to unlearn and reimagine many of our comfortable and predictable ways of working, recalibrate our approach to risk and how we define return on investment (ROI). We must dissolve team and organizational boundaries, developing more sophisticated, partnership-focused collaboration skills with customers, employees, stakeholders, suppliers and even competitors. Developing and nurturing organizational sustainable innovation ecosystems are essential for sustainable innovation success. There is also much to be learnt by looking outwards and across at how other sectors are tackling these challenges, to avoid their mistakes and use successes from other industries to accelerate our own. Different technical sectors have much more in common than it may first appear.

These are the reasons for writing this book. I have been working with organizations to help them overcome any of these challenges and seen the solutions that they have been innovating themselves. By sharing this integrated blueprint for sustainable innovation in technical environments, I hope to help you, as a leader, adapt your team, projects and ways of working to accelerate and de-risk the giant strides forward that we all must take into new territory.

Leading Sustainable Innovation provides a guide on how to deliver eco-innovation within technical environments. It is tailored to meet the needs of innovation leaders and managers working in fields such as transport, engineering, infrastructure, energy, utilities and sciences.

Addressing the unique challenges and opportunities faced by these sectors, this book offers practical methodologies, tools, frameworks and actionable steps that you can implement to create lasting sustainable change for your projects and programmes. Through following a step-by-step process, you will craft a comprehensive roadmap for sustainable innovation, customized for your team or organization. This book aims to empower you, as a technical professional, to drive sustainable growth and resilience in an ever-evolving global landscape and equip you with the knowledge to adapt to changing market demands.

Leading Sustainable Innovation examines multiple aspects of sustainable innovation, such as innovation strategies, state-of-the-art technologies, circular solutions and organizational factors necessary for success. It emphasizes distinguishing good ideas from weak ones and provides guidance on building a sustainable innovation culture. The book underscores the significance of a sustainable innovation ecosystem and multidisciplinary collaboration, spotlighting the need for updated decision-making, governance and project management approaches. It features real-world, global examples and case studies such as Microsoft's Sustainable Datacenters, Maersk, Equinor, the Roads and Transport Authority Dubai, the UK's National Nuclear Laboratory, Northumbrian Water Group, Wunsiedel, Clean Path New York, Port of Tyne, the UK's national 2050 Maritime Innovation Hub, Forestry England and Agriphotovoltaic Assets, to name a few, providing inspiration and transferable, valuable lessons from adjacent industries.

Please download the free resources that go with this book at www.koganpage.com/LSI and do get in touch to let me know how you're doing with your sustainable innovation projects. I promise a personal reply, and you can reach me at jo@bigbangpartnership.co.uk. Thank you for reading. I wish you all the best with your sustainable innovation endeavours.

ACKNOWLEDGEMENTS

I am deeply grateful to a host of individuals and organizations whose support has been invaluable in the creation of this book, which has been a genuine pleasure to write.

First, I would like to extend my heartfelt thanks to my husband John and daughter Amaya for their belief in me and my ideas. Without their encouragement I would not have backed myself and submitted my proposal to Kogan Page. Their ongoing support throughout the writing and editing process inspired me and fuelled my enthusiasm every step of the way.

I am especially grateful to the board and team at Kogan Page for granting me this opportunity and for the substantial support, time and resources that they have dedicated to this book. Having been an avid reader of Kogan Page's publications throughout my academic and professional journey, I am very proud to have become a Kogan Page author. It has been such a privilege and pleasure to develop the concept and ideas for the book with Isabelle Cheng, work on the development edits with Charlie Lynn, the copy-edits with Dawn Cunneen and the typesetting and proofing with Deborah Maloney. My thanks go to the design team for creating a cover that communicates the contents of the book so well, and to the marketing team for their input on the book's value proposition.

I would also like to express my appreciation for the invaluable contributions from the leaders of sustainable innovation who gave their time and expertise to be interviewed for case studies: Matt Beeton and Mark Stoner from Port of Tyne; Tim Whitworth and Paul Knight from National Nuclear Laboratory; Tom Nightingale from Equinor; Frank Allison from FIS360, and the Wild Kielder project team from Forestry England. Thank you, too, to Louise Hunter and her colleagues at Northumbrian Water for their case-study review and support.

Finally, I would like to acknowledge the practical help and encouragement I have received from Karen Drum, who reviewed my early ideas and acted as a great sounding board; Paul Slater, who meticulously helped me to double-check my case-study endnotes; and Gerard van der Burg for pointing me in the right direction for some of my data centre references.

I have been genuinely blown away by how much support I have had from so many friends and colleagues in the process of creating *Leading Sustainable Innovation*. Sincerely, thank you.

01

Introduction to sustainable innovation and its importance

CHAPTER OVERVIEW

We are in the midst of a significant industrial transition. The many decisions we make today will shape our success in securing a sustainable future. *Leading Sustainable Innovation* offers a roadmap for developing solutions that minimize negative impacts and positively contribute to society and the environment.

This first chapter sets the stage for *Leading Sustainable Innovation* and provides an overview of its step-by-step approach. It features an introduction to, and definition of, sustainable innovation and its importance before delving into some of the challenges of driving sustainable innovation in complex technical industries. The chapter also sets out an overview of the structure and content of the book, highlighting how each chapter builds on the one before, creating a comprehensive, practical roadmap to sustainable innovation.

Introduction

Leading Sustainable Innovation provides a guide on how to deliver sustainable innovation within technical environments. It is tailored to meet the needs of innovation leaders and managers working in fields such as transport, engineering, infrastructure, energy, utilities and sciences. Addressing the unique challenges and opportunities faced by these sectors, *Leading Sustainable Innovation* offers practical methodologies, tools, frameworks and actionable steps that you can implement to create lasting sustainable change for your projects and programmes. By following the step-by-step process laid out in the chapters of this book, you'll create a comprehensive roadmap for sustainable innovation, customized for your team or organization. The aim of this book is to empower you, as a leader of technical programmes, to drive sustainable growth and resilience in this ever-evolving global landscape and equip you with additional knowledge to adapt to changing market demands.

The context of *Leading Sustainable Innovation*

There has been a visible increase in the urgency and acceleration of businesses' commitment and actions to combat climate change, despite world economic and political challenges. Industry is in the process of unlearning and reimagining how it innovates, with consideration of legacy and impact on future generations being more important now than ever before.

In addition to creating new solutions, industry is charged with updating older, existing assets with today's sustainability goals in mind. I have seen these changes first-hand in my day-to-day work in the maritime, energy, utilities, construction, technology, science and other sectors.

Maritime 2050,[1] the UK Government's strategy for the maritime industry, was published in 2019. It includes decarbonization and net zero targets. Since then, particularly through my work setting up and leading the UK's 2050 Maritime Innovation Hub, I've seen the active pursuit of decarbonization and sustainability in the industry intensify and multiply each year.

It's the same in many other industries around the world. Globally, the water and wastewater sector is facing challenges from renewing and replacing substantial quantities of ageing infrastructure to building new facilities for a growing population, more resilient supply and protection from climate change-related severe weather impacts. Traditional design, build and operate methods are no longer the answer. The speed and scale necessary for solutions to be achieved, combined with decades of underinvestment, environmental imperatives and the opportunities brought by big data, artificial intelligence (AI) and new technology have created a unique, game-changing shift in how things have been done for many years.

To say that the capability and growth of digital technology globally are revolutionary is an understatement. Digital technology has huge potential to drive more sustainable solutions for industry. But this creates another challenge. To store, process and manage all our data, hundreds of water- and energy-intensive, heat-generating internet data centres around the world are needed. They require substantial amounts of land. Much of this managed data, known as 'dark data', we don't even use. A 2022 study by Kez et al suggests that the annual global carbon, water and land footprints resulting from storing dark data might approach 5.26 million tons, 41.65 gigalitres and 59.45 square kilometres, respectively.[2]

All assets have a finite life, beyond which it is not economically feasible to operate them. Being responsible for how assets can be decommissioned with sustainability in mind from the specification and design phase onwards is becoming more integral. Innovators globally are working to solve inherited and unplanned decommissioning challenges for nuclear power stations, offshore oil and gas installations and pipelines, and more.

Materials, energy and digital sectors are in early development stages, presenting various possibilities. It can be difficult to know which horse to back. What types of the several options for green fuels will shipbuilders and shipping lines favour in the future, and how might they ensure consistency and resilience of supply? What about the 61,000 vessels across the world,[3] currently operating on oil and diesel, that need to be retrofitted or replaced at a cost of between $5 million and $15 million each[4] to achieve the International Maritime Organization's target of the sector becoming net zero by 2050?[5,6]

These challenges and emerging potential solutions are so new that innovation leaders and managers working in fields such as transport, engineering, infrastructure, energy, utilities and sciences are learning and working things out as they go. The new way of sustainability in technical programmes is still evolving as industries create, test and learn about new materials, technology applications and environmental solutions. We are experiencing a significant industrial transition. The many decisions we make today will inform the course of how successful we are in achieving a more sustainable future.

The complexities of leading sustainable innovation

Leaders of sustainable innovation are currently navigating the tension between the familiar and the novel, embracing the dual nature of progress that is often marked by uncertainty. As conditions evolve, they bring both limitations and fresh opportunities. We face a choice: cling to established practices with predictable outcomes or embrace innovation. The implications of this decision are significant – does our credibility stem from past successes or from our ability to adapt to emerging paradigms? Who has the authority to make these calls?

The times we are in create a clear imperative for innovation, where new approaches emerge to address the shortcomings of our current ways of working, while preparing the ground for a new paradigm. Innovation involves taking a concept and gathering the necessary support to bring it to fruition. For a commercial venture, it's about persuading investors to back a promising enterprise. For a social initiative, it involves drawing support from philanthropic and policy-making sectors to approach a persistent issue with fresh eyes. And for infrastructure projects, it's about venturing to offer new designs, methods and materials, as well as building in resilience for future, adverse climate change impacts.

Yet many innovations won't or just don't take hold. Getting innovation to work doesn't always go to plan. Often businesses don't attract the number of customers they need, pilot programmes fizzle out and ambitious infrastructure projects are forced to make compromises due to time, risk or budget concerns. Today's period of transition is a testing ground. It's driving innovation leaders to pilot new, more sustainable ideas, though only a handful will gain traction and endure. The process varies widely across different sectors and cultures: some systems are nurturing innovation while others are stifling it.

When today's sustainable innovation succeeds, it disrupts the status quo, challenging its dominance and reshaping how things are done. Traditional systems don't always simply vanish but often shift to being supported by innovation. For instance, smart grids integrate digital technology to enhance the traditional electricity grid, optimizing the distribution of power. Similarly, new satellite internet services supplement existing broadband networks, extending coverage to areas previously out of reach.

Successful sustainable innovations embed themselves in society, reshaping power dynamics and forming new interest groups that defend their newly won positions. As these innovations gain broader application, they secure deeper stakeholder support. The momentum of increasing returns then propels them forward, overshadowing other contenders.

Although innovation is commonly associated with the new, it can paradoxically have a bias towards the past. Existing power holders and decision-makers, often with a vested interest in keeping things the same, typically control the resources needed for new ventures. They may attempt to adapt these new ideas to reinforce their own status quo. Governments can also perpetuate this by backing established players, driven by a duty to maintain infrastructure or a reluctance to contest the established order. In some cases, they might uphold the old guard simply because of their entrenched political or financial influence.

Leading Sustainable Innovation aims to provide a guide on how to deliver sustainable innovation within technical environments amid these transitional complexities. It offers practical methodologies, tools, frameworks and actionable steps towards creating lasting sustainable change for your projects and programmes.

Defining sustainable innovation

What is 'innovation'?

Innovation in business terms simply means the process of finding and implementing new ways to overcome problems that are worth solving.

Innovations themselves can be anything from low-risk, quick and easy incremental or sustaining improvements that combine over time to make a big difference, to game-changing disruptions that are transformational breakthroughs that fundamentally reshape a market, industry or sector.

Game-changing innovations redefine rules, standards and practices, giving their adopters a significant competitive advantage. They disrupt the status quo, making previous methods or technologies obsolete and setting a new direction for the future. They're not just improvements; they're paradigm shifts. Electric cars, such as Tesla, are an example of game-changing innovation. Manufacturers introduced a new propulsion system that revolutionized car design, software integration and energy consumption patterns in transportation. The innovation also disrupted infrastructure requirements, driving the need for electric-car-charging facilities at fuel stations and in car parks. In the US, the FAA's ongoing, multibillion-dollar NextGen programme is transforming air travel by shifting from radar to satellite systems and introducing data exchange and automation. Thought to be one of the most ambitious infrastructure projects in US history, NextGen aims to revolutionize the safety, sustainability, efficiency, capacity, predictability and resilience of US aviation.[7]

CHARACTERISTICS OF GAME-CHANGING INNOVATIONS

Game-changing innovations are ones that:

- **Redefine competition:** Game-changing innovations often reset the competitive landscape, rendering older methods or products obsolete and making it challenging for traditional players to compete unless they adapt.
- **Create or transform markets:** They might create entirely new markets or drastically change the landscape of existing ones.
- **Involve novelty:** Game-changers are not just about making something better but about introducing something fundamentally different.
- **Set new standards:** Other businesses often need to adopt similar changes to stay relevant, making the innovation a new benchmark or standard in the industry.

- **Have broad impact:** Game-changing innovations usually have ripple effects, influencing related industries and societal behaviours.

Not all game-changing innovations will share these characteristics, but many will. Only a few organizations succeed in creating and commercializing truly game-changing innovations. Game-changing innovation varies greatly depending on the industry, context and criteria used to define it. Innovation in different fields can have varying impacts and may be considered game-changing in one context but not in another. However, most organizations are being impacted by multiple game-changing innovations and this trend will likely continue as the pace of change shows no signs of slowing down. Businesses need to innovate in response to these game-changers. This could be anything and everything from their behind-the-scenes processes and systems through to their customer offer, or even a full business model redesign and reskilling of staff.

THE REAL CHALLENGES OF INNOVATION

Often, having the great, disruptive idea that will solve the problem is the easiest part. The most challenging aspect of innovation is finding ways to make the solution work in practice. For example, working with existing regulations or lobbying for new ones or finding a solution that works on a small scale in the laboratory but is cost-prohibitive and space-hungry to replicate on a commercially useful scale are all significant challenges. Influencing decision-makers to reframe their view of risk and how to manage it can be another significant hurdle. An example of this is the research and debate on how best to address the safety concerns of autonomous shipping by ensuring continued human oversight, rather than whole reliance on AI.[8]

What is sustainable innovation?

Sustainable innovation has the principle of intergenerational responsibility at its centre, ensuring that new solutions, projects and business models contribute to a thriving economy without compromising the environmental and social systems they depend on.

The term 'sustainability' embodies a collection of important social environmental and economic concepts that feature people and the planet as key stakeholders. It's essential to understand each of these concepts, and their relationships with each other, to fully appreciate what sustainable innovation means in practice.

These topics are explored in the following sections. They include the United Nations Sustainable Development Goals; the terms ESG, CSR and sustainability; greenhouse gases, net zero and decarbonization; and circularity.

UNITED NATIONS SUSTAINABLE DEVELOPMENT GOALS

The United Nations Sustainable Development Goals (UNSDGs) are a universal call to action. They are 17 interlinked goals, established with the ambition of ending poverty, protecting the planet, and generating prosperity and peace for all by 2030. The UNSDGs are intended as a blueprint for collective action and ethical progress.

For leaders of sustainable innovation in organizations, the UNSDGs offer a framework for aligning innovation programmes and projects with broader global priorities. By integrating these goals, organizations can contribute to the collective improvement of pressing world issues through their operations, driving innovation that delivers societal and environmental benefits. Innovating in alignment with the UNSDGs not only enhances corporate responsibility but also has the potential to support profitable growth by demonstrating a culture of purpose-driven progress that resonates with customers, investors and employees. Engaging with the UNSDGs is about more than goodwill. It's a strategic imperative that can propel organizations to becoming known as leaders in sustainable innovation.

Leading sustainable innovation in technical programmes and projects means finding effective solutions to create new assets, as well as update existing ones, that at worst do no harm and at best support improvements to the environment. That is climate change, the carbon cycle, greenhouse gases, the ozone layer, health of the oceans, biodiversity and protection of natural resources. It also means, wherever possible, contributing to the achievement of the UNSDGs. This is especially important around goal 9,[9] which includes 9.1 to develop sustainable, resilient, and inclusive infrastructures and 9.4, to upgrade all industries and infrastructures for sustainability.

Industry, Innovation and Infrastructure (Goal 9)

Sustainable infrastructure and advancements in technology contribute to resilient industries and foster innovation, which are core aspects of this goal.

Sustainable innovation in infrastructure and technical projects can directly contribute to several UNSDGs. The aims and content of this book align directly with the following specific goals.

Affordable and Clean Energy (Goal 7)

Innovative projects aimed at enhancing energy efficiency, increasing the share of renewable energy and improving energy access fall under this goal.

Sustainable Cities and Communities (Goal 11)

Projects that promote sustainable urbanization, such as green buildings and efficient public transport, align with creating inclusive and sustainable cities.

Responsible Consumption and Production (Goal 12)

Innovation that leads to sustainable management and adopts the efficient use of natural resources supports this goal.

Climate Action (Goal 13)

Infrastructure projects that reduce emissions and increase resilience to climate-related hazards contribute to taking urgent action to combat climate change.

Life below Water (Goal 14) **and Life on Land** (Goal 15)

Sustainable projects that protect and restore ecosystems, whether terrestrial or aquatic, align with these goals.

Partnerships for the Goals (Goal 17)

Collaborations in sustainable infrastructure projects can cultivate partnerships to achieve these goals.

Differences between ESG, CSR and sustainability

While ESG (environmental, social and governance), CSR (corporate social responsibility) and sustainability are related concepts that all address various aspects of responsible and ethical business practices, they differ in their scope, focus areas, stakeholder engagement and time frames. ESG is primarily concerned with specific environmental, social and governance factors and is often investor-driven. CSR encompasses a wide range of socially responsible activities and philanthropy. Sustainability takes a holistic, long-term view, aiming to balance environmental, social and economic aspects for a more resilient and harmonious future.

ESG

ESG focuses on three specific areas: environmental, social and governance factors, often used by investors and financial analysts to evaluate a company's performance. It typically involves publishing ESG metrics and performance in sustainability reports, often for investors and regulatory compliance. ESG is often driven by the need to meet investor expectations, reduce risks and enhance financial performance through ESG integration.

CSR

CSR is a broader concept than ESG that encompasses a company's efforts to operate ethically and responsibly, including activities that benefit society, employees and the environment that go beyond financial returns. It includes addressing issues such as philanthropy, community engagement, employee well-being and environmental conservation, among others. CSR activities are usually voluntary, often motivated by ethical considerations, reputation management and a desire to contribute positively to society and the environment.

Sustainability

Sustainability is a comprehensive approach that considers the long-term impact of business operations on the environment, society and economy, aiming for a harmonious balance between these aspects. It encompasses ESG factors as well as broader sustainability dimensions, often requiring a strategic, holistic view. Sustainability considers ecological, social and economic dimensions, often linked to the Triple Bottom Line (TBL) concept (people, planet, profit). Sustainability is motivated by the recognition of the interconnectedness of environmental, social and economic factors, aiming for long-term sustainability and resilience. It takes a long-term perspective, considering the legacy and intergenerational impact of business decisions.

NET ZERO AND DECARBONIZATION EXPLAINED

Greenhouse gases are atmospheric gases, like carbon dioxide and methane, that trap heat from the sun, contributing to the Earth's warming effect. The Paris Agreement is a global accord adopted in 2015, aiming to limit global warming to well below 2°C, preferably 1.5°C, by reducing greenhouse gas emissions and enhancing climate resilience.

Net zero refers to balancing the amount of greenhouse gases emitted with the amount removed from the atmosphere. Achieving net zero means a company or country's activities result in no net impact on the climate from greenhouse gases. Decarbonization refers to the process of reducing carbon dioxide emissions, typically by decreasing reliance on fossil fuels and increasing use of renewable energy sources. It is the strategy for achieving net zero.

The classification of greenhouse gas emissions by an organization is categorized into three distinct 'scopes', each representing different sources and scales of emissions. The term 'scope' originates from the Greenhouse Gas Protocol,[10] the most widely used standard for greenhouse gas accounting globally. This categorization helps in comprehensively understanding and managing the company's carbon footprint.

Identifying and addressing these scopes helps organizations understand and manage their climate impact, a crucial step towards achieving global climate goals.

Scope 1: direct emissions

These are emissions from sources directly owned or controlled by an organization. Examples include emissions from company vehicles and on-site fuel combustion.

Scope 2: indirect emissions from energy

Scope 2 covers indirect emissions from the generation of purchased electricity, heat or steam used by the organization. These emissions occur at the place where the energy is produced, not where it's used.

Scope 3: all other indirect emissions

An organization's value chain is the sequence of activities that it performs to create and deliver a product or service to the market. Scope 3 is the broadest of all the scopes, including all other indirect emissions that occur in an organization's value chain. Examples include emissions associated with business travel, procurement, waste and water usage.

THE DIFFERENCE BETWEEN 'CARBON-NEUTRAL' AND 'NET ZERO'

Although they are often used interchangeably, the terms 'carbon-neutral' and 'net zero' do not mean the same thing. Achieving carbon neutrality involves calculating total greenhouse gas emissions and balancing them by funding an equivalent amount of carbon savings elsewhere. This balancing act often involves investing in renewable energy, tree planting or other projects that reduce emissions in the atmosphere. The idea is to compensate for the emissions that are not being eliminated by making sure that an equal amount of reduction happens elsewhere.

Achieving net zero requires a deeper transformation of practices to ensure that actual emissions reach zero, and any residual emissions are counterbalanced with carbon removal, not just offsetting. Net zero is generally seen as a more ambitious and impactful commitment to fighting climate change, in line with global efforts to limit warming in accordance with the Paris Agreement.

CARBON NEGATIVE

As we've seen, attaining net zero carbon emissions means balancing the carbon emitted with the carbon absorbed from the atmosphere. Being carbon negative means going beyond net zero by removing more carbon dioxide from the atmosphere than the organization emits.

SCIENCE-BASED TARGETS (SBTS)

Science-based targets (SBTs) are emissions reduction goals set by businesses that align with the necessary decarbonization levels to meet the Paris Agreement's objective of keeping global warming below 2°C, and preferably 1.5°C, compared to pre-industrial levels. These targets are rooted in scientific understanding and are crucial in shifting towards a low-carbon global economy. They offer a structured approach for organizations to significantly contribute to global climate goals, advancing towards net zero emissions and broader ecological sustainability. Benefits include:

- **Conformity with global climate agendas:** SBTs align corporate emission reduction plans with scientific insights on climate change, aiding in global climate change mitigation efforts.
- **Promoting organizational change:** These targets drive companies to adopt innovative, sustainable operational methods, influencing entire supply chains and setting new industry benchmarks.
- **Pathway to net zero:** SBTs can be integral to achieving net zero emissions, a balance between emitted and absorbed greenhouse gases.
- **Boosting corporate reputation:** Companies adopting SBTs often gain in credibility and recognition for their genuine commitment to environmental impact reduction.
- **Strategic risk mitigation:** SBTs prepare companies for future regulatory shifts, cater to the growing demand for eco-friendly practices and help manage climate-related risks.
- **Encouraging sustainable development:** SBTs promote a forward-thinking approach, which is vital for long-term environmental care and sustainable growth.

CASE STUDY

Working to become carbon negative at Microsoft

Microsoft has committed to moving beyond carbon neutrality towards a carbon-negative stance by 2030, meaning it will remove more carbon than it emits annually. It also aims to remove an amount of carbon equivalent to all its historical emissions since 1975 by 2050.

To address its carbon footprint comprehensively, Microsoft has launched a plan to significantly reduce emissions across three scopes. By 2025, the company aims to power its operations entirely with renewable energy and electrify its vehicle fleet by 2030. It also plans to reduce scope 3 emissions by implementing an internal carbon tax that will cover all business divisions, creating incentives to reduce emissions across the company's value chain.

The company has also introduced a new $1 billion climate innovation fund to accelerate the development of carbon reduction and removal technologies and will integrate carbon reduction criteria into its procurement processes.

Microsoft's initiatives support various UNSDGs, such as UNSDG 6 (Clean Water and Sanitation) with significant water replenishment projects, UNSDG 12 (Responsible Consumption and Production) through significant waste reduction, and UNSDG 15 (Life on Land) by conserving more land than the organization utilizes.

Key to the strategy is Microsoft's adoption of science-based targets (SBTs), taking responsibility for its carbon footprint and investing in technology for carbon reduction and removal. The company emphasizes empowering customers and suppliers worldwide to reduce their carbon footprints through technology, maintaining transparency in progress through annual reporting and advocating for public policies that support carbon reduction and removal.[11, 12]

CIRCULARITY

Circularity, or the circular economy, is a model of production and consumption that involves sharing, reusing, repairing, refurbishing and recycling existing materials and products for as long as possible. The concept contrasts with the traditional, linear economic model of 'take, make, dispose'.

There are two distinct cycles in the circular economy: the biological cycle, where materials can be safely returned to the environment, and the technical cycle, where products and components are designed to be reused or remanufactured, retaining their highest utility and value.

In the context of sustainable innovation in infrastructure and complex technical projects, circularity is highly relevant because it encourages the design of projects and systems that minimize waste and make the most of resources. For infrastructure, this can mean using materials that are durable, repairable and ultimately recyclable at the end of their life. It can also involve innovative approaches such as modular construction, which allows for components to be repurposed or updated rather than needing full-scale replacement.

Circularity drives sustainability in projects such as these by optimizing resource use, reducing environmental impact and creating economic value through innovative materials and waste reduction strategies. Key drivers of circularity are:

- **Political agendas:** Regulatory policies, legislation and tax implications act as main drivers, urging companies to adopt circular practices to comply with governmental mandates.
- **Resource scarcity:** The limited availability of resources is a pressing concern that drives the need for a circular economy, ensuring the sustainability of both society and the planet.
- **Market demand:** There's a growing demand, especially from governmental bodies and business-to-business sectors, for innovative products and services that adhere to sustainable, circular principles.
- **Product value:** Circular innovation projects have the potential to enhance the value of products, for both consumers and companies, by introducing new, sustainable business models.
- **Cost optimization:** Using fewer materials and increasing efficiency present opportunities for cost savings, especially as taxation on raw materials incentivizes waste reduction.
- **Brand equity:** While challenging to quantify, sustainability efforts in innovation projects can significantly boost a brand's reputation and attract talent.
- **Sector initiatives:** Although currently a minor opportunity, sector-specific initiatives are increasingly being recognized as suitable for co-creation and joint learning, accelerating circularity in various industries.

By embedding circular principles, leaders of sustainable innovation can create solutions and infrastructure that not only serve their immediate purpose but also contribute to a more resilient and regenerative future.

The following are examples of how circular principles can be integrated into sustainable innovation projects.

Design out waste and pollution:

- **Infrastructure design:** Design solutions and systems that minimize waste, in both construction and operation, using renewable or recycled materials.
- **Energy systems:** Implement renewable energy systems that reduce dependency on fossil fuels and decrease pollution.

Keep products and materials in use:

- **Materials:** Using materials that are durable, repairable and recyclable extends the lifespan of infrastructure.
- **Modular construction:** Developing infrastructure with interchangeable parts allows for easy repairs and upgrades without full replacement.

Regenerate natural systems:

- **Green infrastructure:** Projects like living walls, green roofs and urban green spaces restore natural habitats and improve biodiversity.
- **Water management:** Systems that recycle water for multiple uses and replenish local water tables prevent resource depletion.

Rethink business models:

- **Service models:** Transition from product-based to service-based models, where the focus is on the service provided, like shared spaces or mobility as a service.
- **Product as a service (PaaS):** For technical projects, PaaS models can ensure equipment is maintained and recycled responsibly.

Use waste as a resource:

- **Waste-to-energy projects:** Converting waste into energy reduces landfill use and provides a renewable energy source.
- **Circular supplies:** Sourcing materials from waste streams for new projects encourages a closed-loop system.

Drivers of sustainable innovation

Various catalysts are contributing to the rapid shift towards sustainable innovation, spanning from regulatory changes to technological advancements and societal pressures.

- **Changing environmental regulation:** Governments worldwide are continuously evolving their environmental regulations in response to rising concerns about climate change. Strict emission standards, waste management protocols and sustainable construction codes have all necessitated a paradigm shift. Businesses are now driven to innovate sustainably to adhere to these regulations, or risk punitive measures.
- **Government targets for climate change and decarbonization:** As part of their commitment to the Paris Agreement, numerous countries have set ambitious targets for reducing greenhouse gas emissions and promoting a low-carbon economy. This has led to an increased investment and focus on green technology and renewable energy in the infrastructure and energy sectors.
- **Growing capability of digital technology:** Digital technology is increasingly playing a crucial role in promoting sustainability. Advanced tools such as artificial intelligence (AI), the Internet of Things (IoT) and automation can enhance energy efficiency, optimize resource usage and boost waste management efforts. The vast data these technologies provide also assists in tracking and managing the environmental footprint of industries.
- **Community and societal pressures:** Society is increasingly vocal about environmental preservation. This collective voice often

manifests as pressures on industries to adopt eco-friendly practices, thereby driving sustainable innovation. Companies are now more inclined to value corporate social responsibility (CSR) and emphasize sustainable practices.

- **Affordability, ethics, ROI and transparency**: With technological advancements, the cost of implementing sustainable solutions has dramatically decreased. Simultaneously, consumers increasingly favour ethical businesses, making sustainable practices a potent tool for gaining competitive advantage. Return on investment (ROI) from sustainable practices has also proven to be promising, due to long-term cost savings and increased market demand. Lastly, the demand for transparency is driving companies to disclose their environmental impact, encouraging the development of greener solutions.
- **The increasing requirement for sustainability from investors**: Investors are progressively emphasizing environmental, social and governance (ESG) criteria. This shift indicates a preference for companies dedicated to sustainable innovation, driving businesses in the target sectors to innovate with sustainability in mind.
- **The pace of change and uncertainty**: Rapid changes in climate patterns and the associated uncertainties need a more adaptable and resilient infrastructure and energy ecosystem. Consequently, companies are innovating to ensure their operations remain resilient, efficient and viable under fluctuating environmental conditions.
- **Increased government incentives for sustainable solutions**: Governments are offering incentives such as tax benefits, grants and subsidies to encourage businesses to innovate sustainably. These incentives make it economically attractive for companies to invest in green technologies and sustainable practices.
- **The development of innovative new materials**: Materials science is becoming more focused on the creation of eco-friendly and sustainable alternatives to traditional materials. The usage of these materials, from bioplastics to self-healing concrete and

solar-reflective surfaces, is now changing the way infrastructure is built and energy is generated. Use of innovative materials requires a new, more collaborative approach to risk allocation between client and supplier, as factors such as performance over time for materials have not been tested in practice.

However, early-stage technology readiness and adoption levels of new solutions can be slow. The drive for more sustainable solutions is only relatively recent. This means that, for many, technology readiness levels are low to medium, and that several options are present. The result is a 'wait and see' approach unless immediate action is necessary, or early adopters step forward. The challenge can be more systemic. For instance, in shipping, vessel owners are waiting for confirmation of the most used and available green fuels before retrofitting ships or buying new ones, while ports and fuel innovators wait to see which direction vessel owners might take. This situation was referred to as a 'chicken and egg' problem by industry speakers throughout Maritime Innovation Week 2023.

In some contexts, such as product development and distribution for fast-moving consumer goods (FMCG), fashion, software development and the service industry, planning horizons are relatively short and agility is high. But for more complex technical programmes, which are the focus of this book, decisions taken, investments made and actions can last for decades, or even longer. For example, after half a century of nuclear activity, it will take around 100 years to decommission the Sellafield nuclear plant in the UK, and more than 10 times that for its legacy waste to degrade.

UNIFYING PRINCIPLES

There are some unifying principles that characterize the shared challenges of leading sustainable innovation in technical programmes:

- **Resilience and longevity:** Sustainable innovation in technical programmes necessitates the design and development of infrastructures that can withstand the test of time and high-volume usage. They are built to last, robust but also adaptable, able to recover from disturbances and evolve with emerging trends and technologies.

- **Complexity:** The project or programme involves many interconnected components and stakeholders, each with different goals and constraints. Sustainable innovation must address this complexity, integrating all components into a holistic and synergistic system that optimizes resources, reduces waste and enhances efficiency.
- **High investment, expertise and long lead times:** These innovations typically require significant capital investment, specialized expertise and long lead times. Hence, sustainable innovation should focus on leveraging advanced technologies and practices to accelerate processes, lower costs and improve outcomes. Training and knowledge transfer are also crucial to ensure the workforce is equipped with the necessary skills to handle these innovative systems.
- **Uncertain future and rapid change:** The future is increasingly uncertain due to rapid technological change, fluctuating market demands and escalating environmental concerns. Sustainable innovation must be proactive, anticipating potential changes, and incorporate flexibility and adaptability into their design and operation.
- **Risk:** Given the high-stakes nature of these programmes, risk assessment and management are integral to sustainable innovation. Risks related to environmental impact, financial loss, operational failure and social implications need to be identified, quantified and mitigated throughout the infrastructure lifecycle.
- **Investment profile:** Investments in these programmes are often characterized by high upfront costs and long-term returns. Investors seek clear visibility of the potential return on investment (ROI). Sustainable innovation must show how it can deliver wider ROI than traditional financial-only approaches, taking account of factors such as environmental impact, operational efficiency, longevity and societal impact.
- **New supply chain capability:** Supply chains play a crucial role in these programmes. Sustainable innovation requires the creation and integration of new supply chain capabilities, emphasizing

traceability, transparency, resilience and sustainability. This could involve adopting technologies, materials and ways of working.

- **Cost competitiveness and ROI:** Whether in the private or public sector, cost competitiveness and ROI are vital. Sustainable innovations must demonstrate their ability to deliver value for money, reducing operating costs and enhancing performance. For the public sector, it's essential to demonstrate ROI to the taxpayer, showing how investments in sustainable innovation projects deliver benefits to society at large.
- **End of life considerations:** Sustainable innovation involves considering the end of life of infrastructure systems. This includes designing for disassembly and recycling, minimizing waste and mitigating environmental impacts. It also involves planning for decommissioning and ensuring a safe and efficient process that minimizes harm to the environment and society.
- **Retrofitting existing structures and assets:** While much focus is placed on creating new, sustainable infrastructures, an equally important aspect of sustainable innovation lies in the retrofitting of existing structures and assets. Given the vast amount of existing infrastructure, it is crucial to find ways to improve efficiency, reduce environmental impact and extend useful life. This includes integrating renewable energy systems, improving energy efficiency and adapting structures to changing uses and needs. By retrofitting older infrastructure, organizations can achieve significant sustainability gains without the need for complete replacement, providing a cost-effective and resource-efficient solution in many cases. Consequently, sustainable innovation must also embrace the principle of 'renew, not just replace', leveraging cutting-edge technology and design approaches to breathe new life into old infrastructure, creating value where it already exists.
- **Regulatory challenges:** The regulatory environment in these programmes is often complex and stringent. Sustainable innovation must provide for regulatory challenges, ensuring compliance while advocating for regulations that promote sustainability.

A step-by-step process for sustainable innovation: The structure of this book

The step-by-step process in the coming chapters are designed to help you to craft a comprehensive roadmap for sustainable innovation, customized for your programme, team or organization to drive sustainable growth and resilience in this ever-evolving global landscape. Throughout the book you'll find real-world, relevant global examples and case studies to bring the sustainable innovation roadmap to life, and to provide additional inspiration for your sustainable innovation projects.

Chapter 2 gives an overview of the sustainable innovation roadmap that you can create using this book. It provides a step-by-step process for developing an actionable plan that integrates sustainability into complex, high-value technical environments. By the end of the chapter, you'll have started to shape your sustainable innovation roadmap for your project or organization. You'll have:

- Aligned your sustainable innovation goals to your organization's mission, vision and core values.
- Identified which stage of sustainability your organization is at, along with barriers and accelerators to better integration of sustainability.
- Considered a Whole Life Value (WLV) approach to sustainable innovation.
- Reflected on the role of leadership at different levels in achieving sustainable innovation in your own organization and teams.
- Developed a sustainable innovation skills map specifically for your organization or team.
- Begun preparations for your sustainable innovation journey and started to put in place strategies for continuous improvement and long-term success.

The third chapter addresses the dynamic nature of the innovation landscape, focusing on rapid new technology development, societal and regulatory changes, economic pressures and the introduction of early-stage innovations to the market. It provides a guide for effective horizon scanning, with practical content covering:

- Drivers and trends in sustainable innovation.
- The dynamic nature of the innovation landscape.
- The importance of horizon scanning, with a full horizon scanning toolkit that you can use with your team.

In Chapter 4, the focus is on green business models and value propositions for technical industries. Designing and implementing sustainable business models is paramount to drive long-term success and stakeholder value. This chapter explores the principles and practices of sustainable business models tailored specifically for technical industries. Real-world case studies will provide successful examples of business models that align with environmental and societal goals, while driving profitability and competitiveness. Chapter 4 also covers:

- Crafting value propositions that drive competitive advantage and stakeholder value.
- Designing sustainable business models that align with environmental and societal goals.
- Ensuring that infrastructure planning and internal processes align with the sustainable business model.
- Creating contemporary, more appropriate procurement models to distribute risk appropriately and optimize the full lifecycle value.

Consistent with the principles of the circular economy, Chapter 5 discusses strategies for retrofitting aged assets, in addition to approaches for creating new assets that are sustainable from inception. Key themes include:

- The two key types of legacy assets: technology and infrastructure.
- Why legacy assets need a compelling sustainable innovation vision, value proposition and business model.
- How to implement circular economy principles to minimize waste, optimize resource usage and promote closed-loop systems.

- Legacy asset redeployment.
- The challenge of, and strategies for, sustainable innovation with legacy assets.
- Principles for creating new, sustainable, future-ready assets.
- The new skill sets required for innovating legacy assets.

Chapter 6 examines the role of technology and data in facilitating sustainable innovation. It focuses on application within complex technical projects and industries, offering strategies to leverage these tools effectively. Topics covered include:

- Getting the best from people and technology working together.
- Leveraging data analytics and digital transformation to inform sustainable innovation strategies.
- Applying predictive analytics and modelling techniques to drive sustainable growth and efficiency.
- Contemporary approaches to environmental and financial business cases for sustainable innovation.
- How companies can empower their customers to make decisions based on data and analytics, and in turn make their own organizations more efficient and sustainable.

The sustainable innovation challenge – balancing sustainability, affordability, value, time and risk – is the primary subject of Chapter 7. It addresses the multifaceted nature of sustainable innovation. It also delves into strategies for overcoming barriers to sustainable innovation. This chapter addresses each of the five pillars of sustainability, affordability, value, time and risk in turn before exploring frameworks and models that consider all five aspects.

Chapter 8 is dedicated to exploring why an updated approach to decision-making and governance is needed for sustainable innovation, providing insights and strategies for effective approaches within this context. It proposes strategies for effective decision-making and governance in sustainable innovation, and for measuring success via a sustainable innovation KPI dashboard. The chapter also features options for a sustainable innovation governance structure and programme sponsorship model.

Building a sustainable innovation strategy and reimagining the innovation pipeline are at the centre of Chapter 9. It comprises a step-by-step guide on how to develop a sustainable innovation strategy, with a focus on reimagining the innovation pipeline to accommodate the unique demands of sustainable innovation.

Chapter 10 focuses on the importance of creating a sustainable innovation ecosystem, and how to go about it. Having a sustainable innovation ecosystem is critical for fostering collaboration, knowledge exchange and cross-industry partnerships. This chapter explores strategies for building an enabling environment that supports sustainable innovation. From developing systems and policies to engaging diverse stakeholders, this chapter shares insights on how to establish a robust ecosystem for your own organization. Themes also include successful collaboration approaches for working with a complex range of supply chain partners and creating an enabling environment through systems, incentives and supportive policies.

Chapter 11 provides practical guidance and a toolkit for successful ideation and knowledge exchange models for multidisciplinary, expert teams, including:

- Understanding the importance of ideation and knowledge exchange in sustainable innovation.
- Exploring different models for ideation and knowledge exchange.
- Practical steps, with real examples to foster creativity and knowledge exchange within multidisciplinary expert teams.

Chapter 12 presents an innovation project management approach for a new paradigm. This chapter introduces a fresh approach to project management that caters to the specific needs and challenges of sustainable innovation in complex, technical environments. This is rarely a linear process. Project management for future-ready programmes needs to allow for trials, iterations, learning and rapid course correction if needed. Traditional methods are too heavy, cumbersome and staid. Chapter 12 also shares practical tools and methods for managing sustainable innovation projects more successfully, and for combining all the themes in this book to shape your cohesive, practical sustainable innovation roadmap.

Conclusion

The driving forces for sustainable innovation are becoming ever stronger. The challenges of leading significant sustainable innovation projects continue to grow. From the combined impact of factors such as decision-making inertia, perceptions of risk, lack of clarity on which new technology, material or solution to back, and significant skills gaps when it comes to progressing future-ready, sustainable innovation in today's fast-changing world, a new roadmap and accompanying set of insights, tools and approaches are needed. This book aims to provide the roadmap, insights, tools and approaches to help today's leaders of sustainable innovation navigate the new paradigm and be ready as that paradigm itself also changes and evolves.

ACTION CHECKLIST

- Check your understanding of the key sustainable innovation principles and concepts: sustainability, ESG, CSR, net zero and decarbonization, and circularity.
- Consider how the sustainable innovation principles and concepts apply to your leadership context and activities.
- Reflect on how you might level up your organization's, team's or project's sustainable innovation impact effectively, based on these principles.
- Map your sustainable innovation goals against the UNSDGs.
- Identify the key drivers of sustainable innovation specific to your leadership role.
- Pinpoint the factors that accelerate and hinder successful sustainable innovation in your context.

Notes

1 Department for Transport (2019) Maritime 2050: navigating the future, www.gov.uk/government/publications/maritime-2050-navigating-the-future (archived at https://perma.cc/BTL2-BUP4)

2 Al Kez, D, Foley, A M, Laverty, D, Furszyfer Del Rio, D and Sovacool, B (2022) Exploring the sustainability challenges facing digitalization and internet data centers, *Journal of Cleaner Production*, [online] 371, p. 133633, www.sciencedirect.com/science/article/pii/S0959652622032115 (archived at https://perma.cc/Q7VK-EQHL)

3 Department for Transport (2023) Shipping fleet statistics: 2022, www.gov.uk/government/statistics/shipping-fleet-statistics-2022/shipping-fleet-statistics-2022 (archived at https://perma.cc/83LA-H5XR)

4 DNV (2023) Challenging road ahead for retrofitting to dual-fuel engines, www.dnv.com/expert-story/maritime-impact/challenging-road-ahead-for-retrofitting-to-dual-fuel-engine.html (archived at https://perma.cc/T352-MK9A)

5 Balci, G and Surucu-Balci, E. Green fuels in shipping face major challenges for 2050 net zero target, The Conversation, 27 September 2023, www.theconversation.com/green-fuels-in-shipping-face-major-challenges-for-2050-net-zero-target-211797 (archived at https://perma.cc/84JB-CPSS)

6 International Maritime Organization (2023) Revised GHG reduction strategy for global shipping adopted, www.imo.org/en/MediaCentre/PressBriefings/pages/Revised-GHG-reduction-strategy-for-global-shipping-adopted-.aspx (archived at https://perma.cc/WB4T-63TC)

7 Federal Aviation Administration (2023) Next Generation Air Transportation System (NextGen), www.faa.gov/nextgen (archived at https://perma.cc/LNC9-MYNY)

8 Veitch, E and Alsos, O A (2022) A systematic review of human-AI interaction in autonomous ship systems, *Safety Science*, [online] 152, p. 105778, www.sciencedirect.com/science/article/pii/S0925753522001175?via%3Dihub (archived at https://perma.cc/3BDS-T5WL)

9 The Global Goals (n.d.) Industry, innovation and infrastructure, www.globalgoals.org/goals/9-industry-innovation-and-infrastructure/ (archived at https://perma.cc/MJR8-QV7E)

10 Greenhouse Gas Protocol (n.d.) www.ghgprotocol.org/ (archived at https://perma.cc/678N-PF3K)

11 Smith, B. Microsoft will be carbon negative by 2030, Official Microsoft Blog, 16 January 2020, blogs.microsoft.com/blog/2020/01/16/microsoft-will-be-carbon-negative-by-2030/ (archived at https://perma.cc/LQ7C-CNRK)

12 Microsoft. On Science Based Targets Initiative's removal of net zero commitment, Microsoft On the Issues, 14 March 2024, https://blogs.microsoft.com/on-the-issues/2024/03/14/on-science-based-targets-initiatives-removal-of-net-zero-commitment/ (archived at https://perma.cc/95FY-6GEV)

02

Creating a sustainable innovation roadmap

CHAPTER OVERVIEW

The step-by-step process outlined in the coming chapters is designed to help you to craft a comprehensive roadmap for sustainable innovation, customized for your programme, team or organization to drive sustainable growth and resilience in this ever-evolving global landscape.

This chapter demonstrates how to:

- Align your sustainable innovation goals to your organization's purpose, mission, vision and values.
- Identify which stage of sustainability your organization is at, along with barriers and accelerators to better integration of sustainability.
- Connect your sustainable innovation projects to your organizational ambitions through OKRs (Objectives and Key Results) and KPIs (Key Performance Indicators).
- Consider a Whole Life Value (WLV) approach to sustainable innovation.
- Achieve effective leadership at different levels to deliver sustainable innovation across an organization, and in individual teams and projects.
- Begin preparations for a sustainable innovation journey and put strategies in place for continuous improvement and long-term success.

Introducing the sustainable innovation roadmap

A sustainable innovation roadmap points you to where you need to go, and what you need to put in place to get there (see Figure 2.1). It

FIGURE 2.1 Sustainable Innovation Roadmap: overview

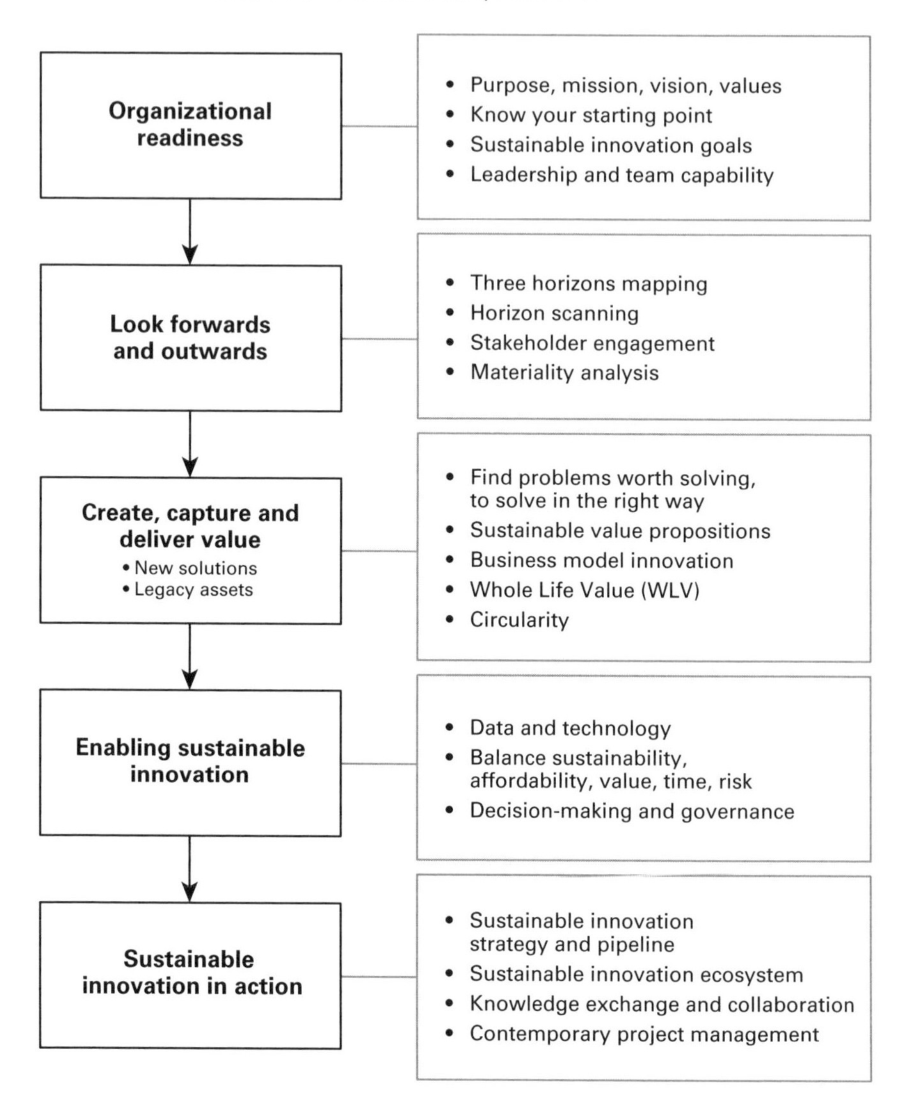

shows how you plan to manage the innovations you'll use to deliver future-ready solutions in your technical programmes and projects. In addition, your sustainable innovation roadmap is the bridge between your organizational vision and strategy and the day-to-day decision-making and actions that will make them a reality. It provides a framework for the organization to work to, and a series of lenses that help you to see your innovations from the different perspectives necessary to achieve the required balance of sustainability, affordability, value creation and risk. A sustainable innovation roadmap will help you to shape new, fit-for-purpose processes and approaches as your organization transforms from traditional ways of working to a new, more relevant, and future-facing model needed for project success, with the agility to evolve as new innovations emerge and the marketplace changes.

As you progress along your Sustainable Innovation Roadmap, you'll find that at times you'll need to revisit your ideas and refine your thinking as you get more specific about your plans, as changes happen, and as new information and insights emerge. You'll also need to move in and out of divergent and convergent thinking phases to harness creativity and analytic decision-making effectively (see Figure 2.2):

Divergent thinking is the phase in which you explore wide-ranging possibilities without immediate judgement or constraints. In this phase, you gather lots of data, and generate a multitude of ideas, considering various perspectives and alternatives.

Convergent thinking, on the other hand, is the phase in which you narrow down these options to feasible and actionable solutions. This phase involves critical thinking and evaluation, selecting the best ideas to move forward with.

In an innovation process, you often start with divergent thinking to open up the problem space and gather a large pool of ideas. As you progress, you switch to convergent thinking to refine these ideas and decide which ones to develop further. This cycle repeats multiple times throughout the process, as new insights and information can send you back into a divergent phase before converging again on a refined set of solutions.

The key is to balance both kinds of thinking: being open and expansive to explore new ideas while also being analytical and discerning to focus efforts on making the best decisions and choices based on the information available at the time.

FIGURE 2.2 Divergent and convergent thinking

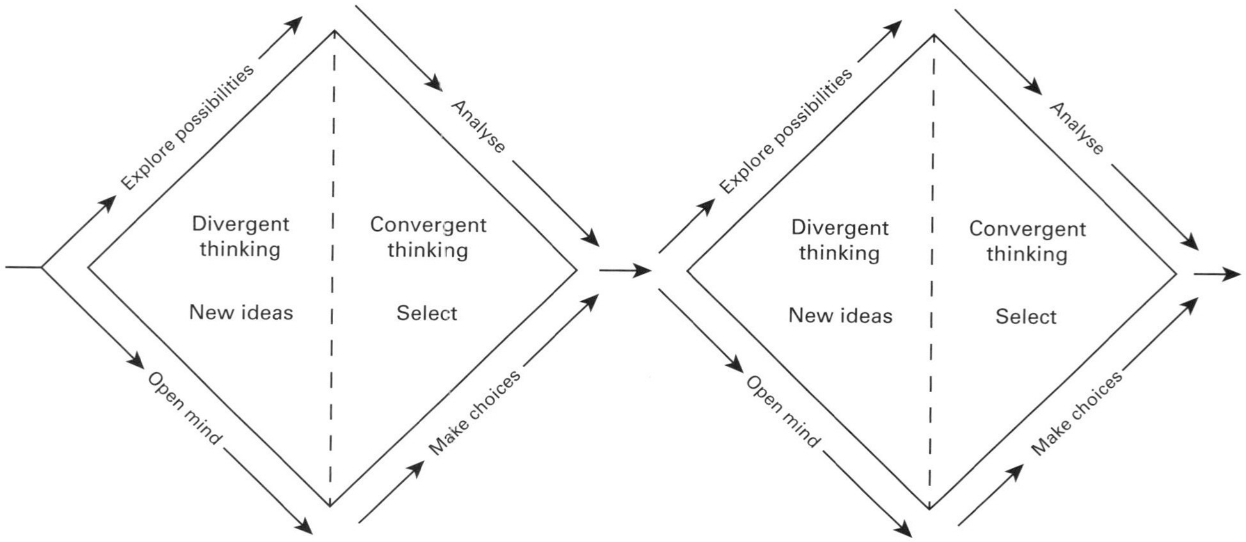

Start with purpose

Organizational purpose is the driving force that defines a company's core reason for being beyond making a profit. It's the guiding star for strategy, operations, decision-making and work culture, informing the ways a company serves its customers, engages with its employees and contributes to the broader community. Having a clear, well-communicated and compelling organizational purpose serves the business and your sustainable innovation projects in the following ways:

- Alignment: Purpose unifies the efforts of employees, offering a common goal that everyone can work towards.
- Motivation: It can foster a greater sense of commitment and motivation among employees, leading to increased productivity and innovation.
- Brand differentiation: A well-articulated purpose can set a company apart from its competitors, highlighting its unique role in the marketplace.
- Resilience: Purpose-driven organizations often exhibit greater resilience in the face of challenges, as their purpose provides a buffer and a reason to persevere.
- Stakeholder trust: Companies that operate with a clear purpose tend to build deeper trust with customers, investors and the community.

Purpose-driven sustainability matters for business performance. Companies with a strong sense of purpose tend to outperform their rivals, securing larger portions of the market and expanding at a rate three times quicker on average, leading to greater employee and customer satisfaction.[1] Purpose-led businesses report 30 per cent higher levels of innovation and 40 per cent higher levels of workforce retention compared to their competitors.[2] The business case, as well as the moral imperative, for purpose-driven, sustainable innovation is strong.

EXAMPLES OF ORGANIZATIONAL PURPOSE

- **Skanska**: One of the world's leading project development and construction groups, Skanska has a purpose that reads 'Building for a better society'.[3] This reflects its commitment to building for the public good and enhancing communities with sustainable and innovative solutions.
- **Ferrovial**: A Spanish multinational company involved in the design, construction, management and operation of transport infrastructure and urban services, Ferrovial states its purpose as: 'To develop and operate sustainable, innovative, and efficient infrastructures creating value for our stakeholders. We want to contribute to making the world more open, connected, and sustainable. We want to develop infrastructures for a world in motion. This is our purpose.'[4]
- **Maersk**: Maersk is a global leader in container logistics and supply chain services, offering solutions that cater to a wide array of customer needs across the entire supply chain. Maersk states its purpose as: 'We are committed to sustainable growth. Our purpose serves as the foundation and compass guiding our work towards a world where global trade distributes economic and social benefits, without negatively impacting individuals, communities, or the environment.'[5]
- **National Nuclear Laboratory**: The National Nuclear Laboratory brings together the UK's nuclear research and development capability into one organization. Its purpose is: 'Nuclear science to benefit society. We are here to help make sure our sector can deliver environmentally and financially affordable solutions to some of the biggest challenges of the twenty-first century.'[6]

Know your starting point: The United Nations Sustainability Stages Model

To achieve your destination, you need to know your starting point so that you can create a roadmap to get there from where you are now.

The United Nations Sustainability Stages Model is a helpful blueprint for organizations to use to support their sustainability journeys.[7] It maps out a five-step process to progress from basic compliance to making sustainability the cornerstone of operations and innovation.

The model provides a practical sequence for improvement, starting with initial responses to sustainability challenges and culminating in a strategic integration of these principles into the company's core mission. It details clear goals and expected results at each stage, ensuring measurable progress. The model also aligns with the United Nations Sustainable Development Goals, as discussed in Chapter 1. Ultimately, the aim is for companies to adopt sustainability as the driving force behind their business decisions, supporting innovation and creating value that benefits both the company and the wider community.

The Sustainability Stages Model integrates financial value drivers – growth, productivity and risk management. The model outlines a progression through five stages of sustainability: starting from crisis management, moving to compliance, resource optimization, market differentiation and, finally, being purpose-driven.[8]

The five stages of the United Nations Sustainability Stages Model

STAGE 1: CRISIS MANAGEMENT

In the crisis management stage, the organization is reactive, trying to keep up with what it needs to do to comply with regulations and customer requirements to preserve its reputation, but no more. Sometimes it fails to comply or comes close to failure. For example, automotive manufacturer Volkswagen confessed to manipulating emissions testing for several of its vehicles. They installed software that could identify test scenarios and alter engine performance to pass emissions standards. While Volkswagen promoted their vehicles as low-emission and environmentally friendly, it was discovered that the actual emissions were up to 40 times over the US legal limit for nitrogen oxide pollutants.[9,10] Volkswagen has worked hard since the 2015 emissions scandal to rebuild brand reputation and demonstrate its commitment to sustainability.[11]

STAGE 2: COMPLIANCE

In the compliance stage, the organization adheres to all required regulations and standards around sustainability but views them as a cost and an unwelcome distraction. The approach to sustainability could be described as transactional. For example, according to Deloitte's 2023 survey on Global Human Capital Trends, a vast majority, 84 per cent of leaders, recognized the importance of grasping how sustainability affects their organizations and the critical need for clearly defined leadership to propel progress and outcomes. This was seen as key to their success. However, only a minority, 21 per cent, felt their organizations were fully prepared to tackle these sustainability challenges.[12]

STAGE 3: RESOURCE OPTIMIZATION

At stage 3, the organization begins to translate its sustainability activities into benefits via resource optimization. Examples of these are increased profitability through cost savings from reduced energy, material usage or revenues from the monetization, reuse or recycling of waste products. The focus on sustainability is enough to enhance the organization's reputation and support employee and stakeholder engagement.

STAGE 4: MARKET DIFFERENTIATION

There is a significant step change between stages 3 and 4. Sustainability is seen as an investment in developing competitive advantage as the organization strives to achieve customer-focused differentiation as a result. For example, in 2023 Radius Recycling (formerly Schnitzer Steel until July that same year), in Portland, Oregon, was named the world's most sustainable business. Despite the carbon-heavy reputation of steel, the company makes most of its revenue from recycling steel and other metals, with a significant portion coming from low-carbon steel produced with hydropower. The company leverages its sustainability developments to stand out in the market. 2023 was one of Radius Recycling's most profitable years since it began operations in 1906.[13,14]

ACHIEVING STAGE 5: PURPOSE-DRIVEN SUSTAINABILITY

As discussed earlier, an organization's purpose is the distinctive contribution that it aims to make to people and the planet through the course of its everyday activities. Being purpose-driven matters because it provides a strong foundation for integrating sustainability into the entire organization, across the value chain – culture, processes, systems, decision-making – in all department teams. An example of an organization that exists to drive sustainability is the Asian Infrastructure Investment Bank (AIIB). To deliver its purpose of developing the 'Infrastructure for Tomorrow' (i4t), AIIB only commits to green and innovative infrastructure projects, ensuring that sustainability, innovation and connectivity are integral to its initiatives.[15]

AIIB's i4t includes the following commitments:

- Environmental: 'Addressing ecological impacts like water and air quality, biodiversity, pollution and climate change.'
- Financial and economic: 'Projects with sound return on investment that raise economic growth and increase productivity.'
- Social: 'Giving inclusive access, particularly to citizens excluded from access to infrastructure services.'[16]

DIAGNOSTIC TOOL FOR ASSESSING UN SUSTAINABILITY STAGES

It's important to carry out an honest, critical assessment of the stage your organization is at, and why, so that you can develop a practical action plan to improve the integration of sustainability into your day-to-day operations and longer-term innovations.

Start by involving your leaders and team in completing the diagnostic tool for assessing your Sustainability Stages. Doing so will help you to pinpoint where you are and track your progress as the organization develops its approach to sustainable innovation.

Stage 1: crisis management

- **Compliance level**: Are regulations and customer requirements met consistently?
- **Reactive measures**: How often does your organization engage in last-minute efforts to address sustainability issues?
- **Reputation management**: Are there instances of sustainability-related reputational challenges?

Stage 2: compliance

- **Regulatory adherence**: Is your organization maintaining consistent compliance with all relevant sustainability regulations?
- **Cost perspective**: Are sustainability efforts viewed solely as costs rather than investments?
- **Transactional approach**: Does your organization view sustainability as simply a mandatory requirement, without aiming for extra value?

Stage 3: resource optimization

- **Cost savings**: Has your organization identified and achieved cost savings through reduced resource usage?
- **Revenue from sustainability**: Is your organization generating revenue from waste valorization, reuse or recycling? (Waste valorization is the process of converting waste materials into more valuable products, reducing environmental impact and promoting sustainability.)
- **Reputation and engagement**: Has your organization enhanced its reputation and stakeholder engagement through sustainability efforts?

Stage 4: market differentiation

- **Investment in sustainability**: Is there a clear investment in sustainability for competitive advantage?
- **Customer-focused differentiation**: Does your organization differentiate itself in the market through sustainability initiatives?
- **Integration into business strategy**: Is sustainability integrated into your organization's business strategy and decision-making?

Stage 5: purpose-driven sustainability

- **Organizational purpose**: Is there a clearly defined organizational purpose that includes sustainability?
- **Value chain integration**: Is sustainability integrated across the entire value chain, including culture, processes, systems and decision-making?
- **Team involvement**: Are all departments and teams actively involved in sustainability efforts in line with the organizational purpose?

The sustainable innovation roadmap in this book will help to accelerate your organization's journey through the stages, embedding sustainable practices into your processes and culture as you go. Make sure that you record your starting point, set some milestones for which stage you want to be at and when, and repeat the exercise periodically to track your progress and reset your goals. It's also useful to test and validate your perceptions with those of your key stakeholders to get a balanced perspective on sustainable innovation priorities, and how well you're really doing.

If your organization is already at stage 5, consider how you might be even more ambitious, as well as how you might work to support other organizations who are further behind on the journey. There is always more that can be done, and complacency is not an option. Firstly, you'll ultimately fall behind as others catch up and eventually overtake you. Plus, by levelling up your sustainability performance, you will discover new ways to grow your business profitably on an ethical basis.

The importance of mission, vision and values in leading sustainable innovation

Defining mission and vision

As we've seen, a company's purpose defines its core intent and the larger aspiration behind its business activities, looking beyond mere profitability to its broader societal and environmental role.

Conversely, a company's mission outlines its operational objectives, detailing the specific services it provides, its target audience and how it seeks success. Purpose represents a firm's legacy in the global narrative, while the mission lays out the tactical roadmap it follows to leave its mark.

A vision is aspirational. It's a statement of where the business, team or person wants to be in the future. A great vision looks ahead, is purposefully ambitious and realistically stretching. A company's vision projects a compelling image of the future it seeks to create, acting as a motivational 'North Star' and charting a path for the organization's future endeavours. This is distinct from the organization's mission, which concentrates on the current landscape, communicating the company's reason for being and its fundamental goals. The mission articulates the strategies and practices the organization employs to achieve its visionary state. The vision communicates the ultimate outcome the organization strives for, whereas the mission describes the practical steps it will take to realize that outcome.

EXAMPLES OF VISION AND MISSION STATEMENTS

Vision

Maersk: 'Our vision is to transform the flow of the foods, goods, data, and materials that sustain people, businesses, and economies the world over. To enable the exchange of ideas, culture, and trust for a truly integrated world where value is created for everyone.'[17]

The Road and Transport Authority, Dubai: The Roads and Transport Authority (RTA) holds the mandate to design and deliver transportation, road and traffic solutions within Dubai, as well as to facilitate connectivity between Dubai, other Emirates in the UAE and neighbouring countries. Its vision statement is to be 'The world leader in seamless and sustainable mobility'.[18]

Clean Path New York: The project represents an $11 billion investment in renewable energy, featuring a combination of wind and solar projects with a capacity of 3,800 MW and a new 175-mile hybrid underground and submarine transmission line. This infrastructure will facilitate the annual generation and delivery of over 7.5 million megawatt-hours (MWh) of clean energy, sufficient to supply power to more than 1.5 million homes in New York. The project is a strategic public–private partnership. Clean Path New York's vision is: 'Our Clean Energy Future is All Renewable. All New York.'[19]

Mission

The Road and Transport Authority, Dubai: 'We provide seamless and safe travel with innovative, sustainable mobility solutions and services to make every journey in Dubai a world-class experience.'[20]

Clean Path New York: 'Clean Path New York is focused on solving complex challenges and meeting New York State's ambitious climate goals.'[21]

Department of Infrastructure, Transport, Regional Development, Communications and the Arts, Government of Australia: 'Strategic leadership and coordination in supporting the safe and legal deployment of future transport technologies to enhance and improve transport accessibility, road safety, congestion and productivity for all Australians.'[22]

Organizational values

Values are the important principles, beliefs and ways of behaving and seeing the world that connect everyone. They represent *how* the organization *does* business.[23] This works on an organizational level, as well as an individual innovation level. It's important to make sure that your innovation activities align with your organizational goals.

EXAMPLES OF ORGANIZATIONAL VALUES

Maersk:

Constant care

Take care of today, actively prepare for tomorrow. Whether solving today's challenges or exploring opportunities to shape the future, we anticipate, innovate, and strive to improve everything we do.

Humbleness

Listen, learn, and share to create value for others. We stay curious, open-minded and respect other perspectives, always seeking to learn from each other, our customers, and the world around us. We only succeed together.

Our employees

The right environment for our people. Connected by real purpose, we create opportunities to grow, develop, and exceed expectations. We win together as a diverse and global workplace where people feel safe, valued, and empowered.

Uprightness

Our word is our bond. Every day, we earn the trust of our customers and partners. They can rely on us to keep our promises and do the right thing, even when it's hard. We speak openly and honestly, and always act with integrity.

Our name

Everything we stand for. Our name is a promise and a commitment to trust and excellence. We are all ambassadors representing and safeguarding the Maersk name, striving for a more sustainable and integrated world.[24]

The Road and Transport Authority, Dubai:

Preserve reputation

Safeguarding our corporate reputation as a trustworthy, transparent, and reliable partner.

Strive for pioneering

Aiming for world-class standards in all regards.

Leadership and teamwork

Taking initiative and ownership and building an environment of trust and collaboration to achieve our goals together.

Promote tolerance and collaboration

Actively building and protecting a culture of support, collegiality, and respect for every employee of RTA.

Be pioneering and challenge conventional thinking

Relentlessly challenging the status quo, conducive to positive change and taking bold future bets.[25]

If your organization's purpose, mission, vision and values were created some time ago, it's worth reviewing and potentially updating them to make sure that they reflect your sustainability ambitions. This matters because they are the 'North Star' of everything your organization does, from how people work together, what they prioritize, through to being the basis of every decision that you make.

To shape an organization where purpose, mission, vision and values converge into a clear, holistic and sustainable growth strategy, it's important to embed these elements into every facet of the operation. Making them live and breathe in every action and decision taken is key. Authenticity is the foundation. It ensures that actions and decisions resonate with the core statements, building trust internally and externally. When an organization's claims of sustainability or social responsibility are visibly mirrored in its practices, the integrity of the purpose is upheld, compelling stakeholders to commit and contribute to the collective vision. This integrity translates into enhanced performance, as a workforce united by authentic, clear and lived values moves together towards common goals.

The focus on alignment with purpose, mission, vision and values often has a strong internal emphasis. However, more and more organizations are now ensuring, through their procurement and contract management processes, that their supply chain and key partners share fundamental values and principles, not only regarding sustainability but also more widely.

CASE STUDY

Leading sustainable innovation in practice: Microsoft's Moonshots

Microsoft's Moonshots show how mission, purpose, values and goals drive actions and decisions across the global organization.

Microsoft's mission is:

> To empower every person and every organisation on the planet to achieve more.[26]

Its corporate values are:[27]

> **Respect**
>
> We recognise that the thoughts, feelings and backgrounds of others are as important as our own.
>
> **Integrity**
>
> We are honest, ethical and trustworthy.

Accountability

We accept full responsibility for our decisions, actions and results.

The company refers to their 'values in action' as being:

- **Innovation**, seeing technology as a force for good
- **Diversity and Inclusion**
- **Corporate Social Responsibility**
- **AI** to help people and organizations achieve more
- **Trustworthy Computing**, 'secure, private and reliable computing experiences based on sound business practices'.

Moonshots aim to address a huge problem by committing to innovate solutions that can create significant impact if successful. Microsoft's moonshots are to become carbon negative by 2030, remove its historical carbon emissions by 2050 and establish a $1 billion climate innovation fund. They also include achieving zero waste, becoming water positive (i.e. replenishing more water than it consumes) and restoring more land than the company uses.[28,29]

These goals drive the plans, actions, behaviours and decision-making across the company globally. It includes those for Microsoft data centres.[30,31] Some of the actions taken to achieve the goals include:

- Using 100 per cent renewable energy by 2025.[32]
- Introducing a carbon fee internally for all its projects.
- Developing software to help reduce energy consumption.
- Investing in cutting-edge scientific research on sustainability solutions globally.
- Redesigning business processes to reduce usage of energy and other resources.
- Building infrastructure designed with sustainability in mind.
- Pioneering waste-to-energy projects, such as the DataPlant innovation project in Cheyenne, Wyoming, that is powered by a wastewater treatment plant.[33,34]

Each generation of their data centre design yields major gains in efficiency and sustainability.

CASE STUDY

Extending purpose, mission, vision and values to contractors in the water industry

Introduction to Northumbrian Water

Northumbrian Water Group Limited, owned by CK Hutchison Holdings Ltd, CK Asset Holding Ltd and KKR & Co Inc,[35] Hong Kong and New York stock exchange listed companies respectively, plays a crucial role in the UK's water and sewerage sector. The company focuses on providing high-quality water services in the Northumbrian Water and Essex and Suffolk Water regions, and sewerage services in the Northumbrian Water region. Core to Northumbrian Water's operations is a strong commitment to sustainability, which involves protecting and enhancing the environment, meeting customer needs effectively and positively impacting the communities they serve. Their business activities centre around delivering reliable and resilient services essential for public health, valuing natural capital and ecosystems, ensuring exceptional customer experiences and contributing more broadly to society.

Vision and values

Northumbrian Water's vision is 'to be the national leader in the provision of sustainable water and wastewater services'.

Its values are:

- Customer-focused: We aim to exceed the expectations of our external and internal customers.
- Results-driven: We take personal responsibility for achieving excellent business results.
- Ethical: We are open and honest and meet our commitments with a responsible approach to the environment and our communities.
- Innovative: We continuously strive for innovative and better ways to deliver our business.
- One team: We work together consistently, promoting cooperation and mutual support to achieve our corporate objectives.[36]

Asset Management Periods

AMP8 refers to the eighth Asset Management Period in the UK's water industry. Asset Management Periods (AMPs), typically spanning five years, are phases during which water companies plan and implement their investments and improvements in infrastructure, like water treatment plants and sewage systems. Each AMP is regulated by Ofwat, the Water Services Regulation Authority in the UK, which sets out expectations and targets for water companies. These targets include improving service quality, enhancing environmental sustainability and maintaining affordable prices for consumers.

Working with contractors and other stakeholders

In the context of Northumbrian Water's AMP8, it represents the company's strategic plan and investment framework for a specific five-year period, focusing on upgrading and maintaining their water and sewage treatment infrastructure to meet regulatory standards and customer needs. This involves substantial investment, planning and coordination with contractors and other stakeholders to ensure efficient, sustainable and effective water management and services.

A new 'enterprise' model for working with contractors

Since 2022, the company has been gearing up for its major £4.5 billion AMP8 project, planned for 2025–2030. It invited contractors to participate in its procurement planning for the launch. Northumbrian Water aims to revolutionize its capital investment approach by adopting an 'enterprise' model that emphasizes collaboration, shared values and joint incentives with contractors.[37] The focus is on engaging design and construction firms experienced in collaborative models, emphasizing value delivery, cost-effective and timely outcomes, low carbon initiatives, biodiversity and workforce skill development. Feedback from potential partners shaped the framework's structure, aligned with Northumbrian Water's commitment to sustainability and customer value enhancement.

Behavioural assessment centres

Behavioural assessment centres are a useful way for companies to ensure that the suppliers they select operate in alignment with the company's purpose, mission, vision and values. A behavioural assessment centre is a structured evaluation process. Unlike traditional procurement processes that may focus primarily on price, quality and delivery, a behavioural assessment centre looks at how a supplier's practices, culture and ethos match with the hiring company's core principles.

Key elements of this process might include:

- **Scenario-based exercises**: Suppliers may be given hypothetical scenarios to see how they would respond in situations that test their alignment with the hiring company's values.

- **Interviews and discussions**: Engaging in-depth conversations to understand the suppliers' ethical standards, sustainability practices and corporate culture.
- **Review of past performance and practices**: Evaluating suppliers' history in terms of ethical business practices, environmental sustainability, social responsibility and other values important to the company.
- **Stakeholder feedback**: Gathering insights from various stakeholders within the company to gauge the potential supplier's fit.

This approach helps to make sure that the selected suppliers do not just meet technical and financial criteria but also contribute positively to the company's broader goals, such as sustainability, corporate social responsibility and ethical practices.

Northumbrian Water introduced behavioural assessment centres to its contractor selection process to ensure that the firms they chose to work with demonstrate values and principles that align with theirs.

Sustainable innovation goal setting: the bridge between purpose and action

Goals translate big ambition into attainable targets, setting the agenda for what the organization aims to achieve in the foreseeable future. Sustainable innovation goals are the concrete milestones that chart the course to realizing purpose. They also create a bridge to the mission and vision, guiding the selection of actions to achieve them. With clear goals in place, the organization can implement plans, processes, systems and resources to accomplish these aims, effectively operationalizing its purpose, mission and vision in ways that are consistent with its sustainable innovation values.

Objectives and key results: measurable steps

Objectives and key results (OKRs) break down goals into smaller, measurable outcomes (see Figure 2.4). This stage is where ambition meets reality, defining the precise results needed to consider each goal met. It's a checkpoint for organizational focus, ensuring every team knows what success looks like.

FIGURE 2.3 Sustainable Innovation Roadmap: strategy pyramid

KPIs/targets: gauging progress

Key performance indicators (KPIs) and targets are the metrics that track progress towards objectives. They are the quantifiable signposts that tell us how close the organization is to its destination. These indicators are critical for maintaining momentum and adjusting strategies as needed.

OKRs and KPIs serve distinct but complementary roles in goal setting, especially within the framework of complex sustainable innovation projects. OKRs are a goal-setting framework that pairs qualitative objectives with measurable key results. They are set with the intention to push boundaries and create momentum within an organization. The objective is the qualitative goal set by the organization. The key results are the measurable outcomes that will indicate achievement of this goal. KPIs are then used to monitor the ongoing performance of these projects, providing sponsors and governance bodies with quantifiable data on whether the projects are aligned with the sustainable innovation goals of the organization and are progressing as planned. This ongoing performance tracking is essential for maintaining control over the sustainable innovation project, ensuring resources are being used effectively and supporting informed decisions about future actions.

EXAMPLE OKRs AND KPIs

Objective: Integrate sustainability into infrastructure development practices to enhance environmental stewardship and social responsibility.

Key Results:

- **Reduce carbon footprint**: Lower greenhouse gas emissions from construction activities by 20 per cent within the next 12 months through the adoption of low-carbon technologies and materials.
- **Increase sustainable sourcing**: Ensure that at least 50 per cent of all construction materials used are sourced from certified sustainable suppliers by the end of the fiscal year.
- **Enhance community engagement:** Conduct monthly sustainability workshops with local communities affected by the projects to garner feedback and improve social impact strategies, aiming for a 90 per cent positive feedback rate.
- **Implement renewable energy solutions**: Integrate renewable energy sources, such as solar panels, into 30 per cent of the projects to power site operations by the next two quarters.
- **Water conservation**: Achieve a 25 per cent reduction in water usage across all projects through the implementation of water recycling and conservation systems within the next six months.
- **Increase biodiversity**: Incorporate biodiversity action plans in 60 per cent of projects, aiming to not only minimize harm but also provide net-positive contributions to local ecosystems within the year.

These key results are designed to be SMART (Specific, Measurable, Achievable, Realistic and Time-bound) and directly support the overarching objective of sustainability in infrastructure development.

KPIs should be designed to enable measurement of progress and performance towards each OKR on a weekly, monthly, quarterly or six-monthly basis, depending on the rate of change. The greater the rate of change, the more frequent the measure needs to be.

1. **Carbon footprint reduction KPI**: Percentage reduction in total greenhouse gas emissions from construction operations compared to the previous year.
2. **Sustainable sourcing KPI**: Percentage of construction materials purchased from certified sustainable sources against the total materials used.
3. **Community engagement KPI**: Positive feedback from community workshops through surveys and feedback forms to quantify stakeholder satisfaction.
4. **Renewable energy integration KPI**: Percentage of energy consumption on construction sites that is met by renewable sources.
5. **Water usage KPI**: Volume of water conserved through recycling and conservation systems, compared to the baseline water usage of the previous year.
6. **Biodiversity enhancement KPI**: Number of projects that have successfully implemented biodiversity action plans, measured against the annual target.

Each of these KPIs is aligned with a specific key result from the OKR and provides a quantifiable means of tracking the success of the sustainability initiatives implemented in the organization's infrastructure projects.

Of course, before you set your KPIs you'll need to ensure that they are meaningful by analysing baseline data for your current position and performance history for each OKR.

Bringing strategy to life

The strategy pyramid in Figure 2.3, with your specific OKRs and KPIs above (see Figure 2.4), create a structured approach to working towards your organization's purpose and vision. They are the skeleton that supports your organization's activities and decisions but are not enough on their own, as they only exist on paper. Running vertically in the strategy pyramid (Figure 2.3), both up and down, from purpose to daily actions and behaviours, are values and high-performance collaboration. You need to make sure that your team's behaviours and actions

FIGURE 2.4 Sustainable Innovation Roadmap: goals, OKRs and KPIs

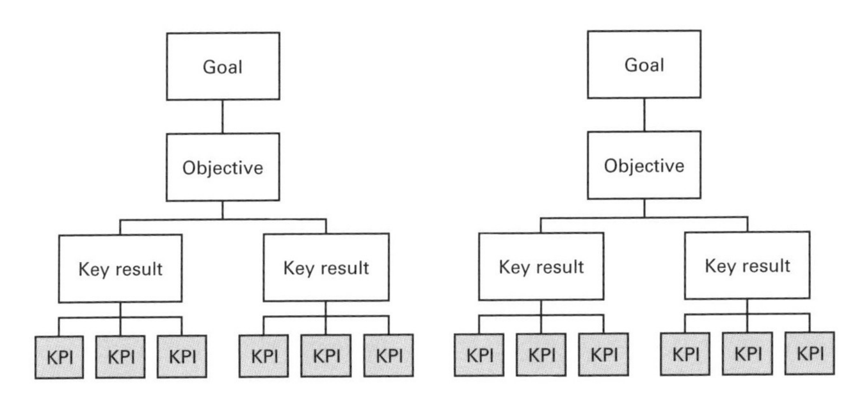

are completely aligned to your organization's purpose and vision, and that your team works together effectively to achieve its goals.

Behaviours and actions represent the everyday efforts and cultural practices of each person in the organization, individually and collectively. They are the outwardly visible manifestation of your values in action. Behaviours and actions include the steps that employees, suppliers and partners take daily and are rewarded for. They are instrumental in achieving your KPIs in line with your organization's purpose and vision because they influence how you do business.

High-performance collaboration is the commitment to working synergistically, leveraging diverse talents and ideas to achieve common goals. We'll discuss collaboration in more depth in Chapters 10 and 11.

Taking a strategic approach to creating sustainable value: materiality analysis

It's all too easy to get bogged down in irrelevant details, or focused on micro-actions that distract attention and effort rather than moving the needle. Prioritization is key, as is a balance between achieving quick wins alongside developing projects that will deliver financial and non-financial payoffs in the medium to long term. Materiality analysis helps to ensure that your organization focuses on those sustainable innovation programmes that will make the most positive impact in working towards its purpose, mission and vision.

Materiality analysis is a strategic process used to identify and prioritize the most significant sustainability issues that are of interest to an organization's operations and stakeholders. It helps to determine which aspects of a sustainability are most impactful and specifically relevant to the organization's innovation programme.

The process involves assessing various factors such as environmental impact, resource usage, ethical practices and social responsibility, within the context of the organization's technical activities and innovations. Doing so means that the organization can focus its efforts and resources on areas where they can make the most meaningful and effective contributions to sustainability, while also addressing the concerns and expectations of stakeholders like customers, employees, investors and the community.

Through materiality analysis, leaders in technical sectors can integrate sustainability more effectively into their innovation processes, ensuring that their projects not only advance technological progress but also contribute positively to environmental and social goals. It's essential to do this on two levels:

1 How material is each sustainable innovation project to your organization as a whole?
2 What aspects of each of these projects are more material than others?

FIGURE 2.5 Sustainable Innovation Roadmap: materiality matrix

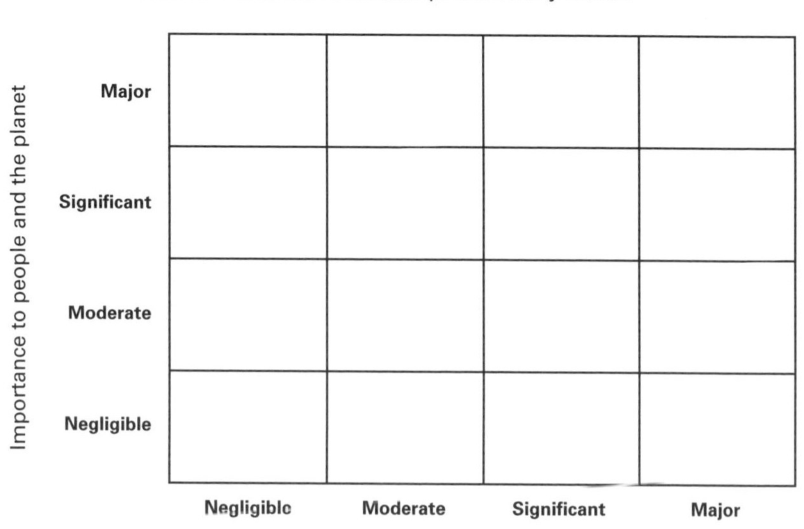

Create a matrix, with the x-axis relating to the importance of each aspect of sustainability to the organization and the y-axis showing the importance of it to people and the planet, as shown in Figure 2.5. Plot your plans and issues onto this matrix. Use it to help prioritize the issues and opportunities that will have the most impact on your organization and its stakeholders.

The dynamic interplay between materiality and value in sustainable innovation

Value is the level of importance, usefulness or worth ascribed to sustainable innovation. It can be quantifiable, such as an investment cost, return on investment figure or growth in market share. In the context of sustainable innovation, value often transcends the tangible and quantifiable, to include perceptions and emotions that resonate with stakeholders. Value is not only about the economic benefit but includes the ethical, environmental and social implications that people experience and believe to be significant.

The perceived or felt value of sustainable innovation can be as impactful as its financial and measurable value. It captures how individuals and communities feel about the innovation's contribution to the greater good, influencing consumer choices, investor decisions and the overall reputation of a company. This kind of value creates a powerful narrative that can drive brand loyalty, attract talent and secure a competitive edge in the market.

WUNSIEDEL: CLEAN ENERGY ON A REGIONAL SCALE

An example of the importance of perceived or felt value is the Wunsiedel regional clean energy project.

Wunsiedel is a town in northern Bavaria, Germany, that has partnered with technology company Siemens to sustainably innovate clean energy production for the local region. Before the clean energy programme, inhabitants were leaving the area due to the decline in the formerly booming porcelain industry.

Former Mayor Karl-Willi Beck and Marco Krasser, CEO of SWW Wunsiedel GmbH, the municipal utility, developed a blueprint for a smart town powered exclusively by renewable energy sources. Their plan led to an all-encompassing infrastructure initiative encompassing educational facilities, leisure amenities, flood defences and, chiefly, their renewable energy network, the WUNsiedler Weg. The plan involved high levels of community involvement and participation.

Wunsiedel is now back to being a thriving community, with residents demonstrating a strong sense of pride and empowerment when they speak to the media about what they have worked together to achieve for their region.[38]

Materiality analysis helps to identify which sustainable innovation projects, and which aspects of those projects, are deemed to be of most value to an organization and its stakeholders. Conversely, the perceived value of different aspects of sustainable innovation can influence what is considered material.

Changes in what stakeholders value can shift materiality assessments. For example, increasing concern about climate change has elevated environmental sustainability as a material issue for many organizations.

The interplay between value and materiality is dynamic. As stakeholder expectations and market contexts evolve, what is considered valuable and what is material can change, meaning that organizations need to continually reassess and adapt their strategies.

ROAD AND TRANSPORT AUTHORITY, DUBAI: MATERIALITY ANALYSIS

In 2021 and 2022 the Road and Transport Authority, Dubai, published their materiality assessments for each year.[39,40] The organization illustrated each year's priorities by plotting key themes onto a materiality matrix for each year as follows:

- The x-axis showing the significance of economic, environment and social impacts.

- The y-axis showing the influence of each issue on stakeholders' assessment and decisions.

The result was the ranking of themes into moderate, material and highly material categories.

These themes changed from 2021 to 2022, demonstrating the importance of keeping materiality analysis up to date as priorities change. In 2021, the top five highly material themes were, in order of importance, occupational health and safety, energy and emissions, people happiness, local community and waste. In 2022, the top five were energy and emissions, occupational health and safety, innovation, people happiness and customer privacy.

HUSQVARNA: MATERIALITY ANALYSIS

Husqvarna is a leading global company, headquartered in Sweden, known for its high-quality outdoor power products. The company's Sustainovate programme is an initiative focused on integrating sustainability into its core business strategy. It aims to reduce environmental impact through innovation, efficient resource use and responsible practices.

Husqvarna published its materiality analysis in its 2022 Sustainovate report.[41]

- The x-axis shows the impact of issues on the business, described as: 'Husqvarna Group's impact on people and the planet combined with the topic's influence on our ability to create value.'
- The y-axis shows stakeholder insights, defined as: 'The degree topics influence stakeholders' perceptions of Husqvarna Group and its ability to create value for them.'

The results shown in the matrix were validated by 360 people, including Husqvarna group management, representatives from stakeholder groups and employees.

Themes were prioritized into the following categories:

- Strategic enablers
- Market leadership
- Trust builders

The business saw some changes from the priorities it had identified in the previous year, illustrating the importance of updating materiality analysis on a regular basis.

Understanding Whole Life Cost, Whole Life Value and Value Engineering

To build on the theme of value and materiality, how sustainable innovation projects are financially appraised can affect their perceived level of impact and importance. It is essential to adopt a more holistic, longer-term mindset when pursuing complex sustainable innovation in technical environments and consider a 'whole life' approach when deciding which activities will best help your organization to achieve its purpose, vision and mission.

Whole Life Cost (WLC) goes beyond a standard project costing approach. The latter considers how much the project will cost to design and build. WLC, on the other hand, accounts for the financial costs of the full lifecycle of a project, through the design, build, operate, decommissioning and disposal phases.

Whole Life Value (WLV) goes further than WLC. It encompasses the financial and non-financial costs and benefits of an innovation, with an emphasis on the inclusion of sustainability factors, such as social and environmental impact, throughout the lifecycle.

As we'll discuss further in Chapter 3, it's difficult to anticipate the future needs and environmental standards over the extended timelines of longer-term sustainable innovation projects. These projects face sustainability challenges because the rapid pace of technological changes can make current sustainable solutions obsolete quickly. These projects also need to deal with fluctuating availability and costs of sustainable materials and resources. Taking a WLV approach helps leaders of sustainable innovation projects to consider these challenges and prepare to make changes, expected or otherwise, as needed.

WHOLE LIFE VALUE (WLV) IN CONSTRUCTION

Sectors such as construction have been more focused on WLV as the demand for more and better environmentally conscious solutions has increased. The industry needs to look beyond the upfront costs and evaluate the economic and environmental impact of a building over its entire lifecycle. This includes initial construction, operation, maintenance, demolition or recycling.

Some of the benefits of taking a WLV approach are:

- Buildings that are designed for WLV tend to be more cost-effective over time. They require less maintenance, are more energy-efficient and have lower operating costs.
- Construction with WLV in mind helps minimize waste, reduce energy consumption and decrease harmful emissions.
- Buildings designed to WLV principles often provide better living and working environments. They contribute to the health and comfort of their occupants, which can improve productivity and well-being.

Value engineering is the process of removing anything from your proposed solution that does not contribute in a useful way, or that increases cost or environmental impact without improving performance to a satisfactory level. It involves reducing the project scope or quality standards, without negatively impacting the outcome.

Value engineering has had a bad reputation when it has been used to slash budgets without due consideration of the resulting quality. When done well, however, it's an essential, systematic review of all project features and costs to ensure an excellent result that delivers financial, sustainability and other non-financial objectives.

We will go into more depth on balancing sustainability, affordability, value, time and risk goals, sponsorship, governance and performance management in Chapters 7 and 8. At this stage, however, it's important to note that clarity of purpose, mission, vision and values is invaluable when creating well-thought-out goals, OKRs and KPIs. They'll facilitate the success of sustainable innovation projects, working towards the organization's ambitions and enabling effective performance tracking, sponsorship and governance.

The role of leadership in sustainable innovation

Successful sustainable innovation requires advanced leadership, from Boards and C-suite executives through to frontline project managers, and all people and technical management roles in between. Because organizations need to create a new operating model to deliver sustainable innovation projects in technical environments, it's necessary for leaders to adjust their mindset, roles and approaches accordingly, acquire new skills and enhance existing ones. Sustainable innovation requires a new leadership mindset and competency framework.

BOARDS AND C-SUITE EXECUTIVES

Senior leadership needs to articulate a clear, compelling purpose for sustainable innovation vision, in line with the organization's mission, vision, values and strategy. They role-model the culture and behaviours they want to nurture across the business, encouraging sustainable innovation and learning from failures, promoting cross-functional collaboration and multidisciplinary working.

In the context of sustainable innovation, senior leaders inspire others by demonstrating a growth mindset, acknowledging that no one person has all the answers, and that, in some cases, there is no single 'right' answer to be found, only the best one available, selected through good judgement, from a range of imperfect options.

PROJECT SPONSORS

Project sponsors are accountable for ensuring appropriate project governance delivery of the required value from the sustainable innovation to stakeholders, in line with the approved business case. They play a leading role throughout the project lifecycle in making sure that the project team operates in an environment conducive to success, with the resources, funding and stakeholder support needed. Project sponsors also monitor and mitigate overall project risk.

The most significant competency enhancements for project sponsors of sustainable innovation are reimagining how risk management might be achieved as new or less familiar methods, technologies and materials are used. Creating an environment in which collaborative working and creative problem-solving can flourish is also imperative, as the organization, its partners and supply chain grapple with new problems requiring solutions that achieve much more than financial return and customer satisfaction.

PROJECT MANAGERS

Project managers, of course, are tasked with being on the frontline, working in the everyday reality of making sustainable innovation happen for real. They bring together and coordinate mixed teams of experts from different disciplines to produce results in line with commitments made on time, quality and impact. Their ability to promote and facilitate collaborative working and problem-solving for novel, more sustainable outcomes and navigate unchartered territory can be a steep learning curve.

SUBJECT MATTER EXPERTS (SMEs)

Subject matter experts (SMEs) are leaders too, whether they have a direct team of their own or not. Because they possess deep technical expertise, it is their role to appropriately inform and influence the wider project team on potential solutions and recommended approaches. For sustainable innovation, the skills SMEs need more than ever before are ideation, or idea generation, that is, coming up with new, more sustainable ways to solve problems and opportunities within project constraints, and communication, to influence and lead others who often have different expertise, to support innovative choices and decisions. SMEs also need to keep up to date with advancements in process approaches, materials and technology innovations and future new regulatory requirements in their area of expertise to lead and inform the wider project team.

LEADERSHIP FOR A NEW PARADIGM

There is a myriad of other roles that directly drive or support sustainable innovations, of course, and in Chapter 12 we will dive more deeply into the new project management approach needed for sustainable innovation in technical environments. For now, the key point is to recognize that what got us here will not get us there. Sustainable innovation will only happen successfully if leaders at all levels in organizations equip themselves with the skills and mindset they need to give their optimal contribution to accelerating the adoption of the necessary new paradigm, for people, planet and performance.

Conclusion

Your sustainable innovation roadmap provides a step-by-step process for developing an actionable plan that integrates sustainability into technical environments. It includes the fundamental elements that most organizations need to consider. That said, every organization and project is different, of course, so this sustainable innovation roadmap is designed so that you can adapt it to suit your specific requirements.

Your sustainable innovation roadmap will also help you to:

- Benchmark your current and proposed practices, and find opportunities to improve or enhance them, or maybe even create brand-new, fit-for-purpose processes from scratch.
- Develop or select the most appropriate solutions, processes and approaches for your projects, based on the information available.
- Optimize returns on investments by focusing time, money and attention on using the most appropriate innovative materials, technology and methods for a balanced, future-minded solution.
- Communicate your sustainable innovation plans transparently to internal and external stakeholders, accelerating onboarding of new ways of working across your organization.

- Go beyond small, incremental improvements to your existing approaches and make well-considered step changes.
- Take a more holistic, circular, Whole Life Cost (WLC) and Whole Life Value (WLV) approach to your projects. This allows value engineering of your project to identify appropriate trade-offs, enabling you to shape the most effective, efficient solution to meet your needs.

ACTION CHECKLIST

- Download your free, editable pdf of the Sustainable Innovation Roadmap templates at www.koganpage.com/LSI, ready to complete as you progress through this book. Or you can replicate the templates yourself as they appear in the book.
- Assess which stage of the UN's Sustainability Stages Model your organization is at, and why. Focus on accelerating your journey to becoming fully purpose-driven by creating your sustainable innovation roadmap, using the chapters in this book.
- Write down a clear, compelling purpose statement for your sustainable innovation project(s). Show how your purpose statement aligns with your organization's purpose, mission, vision and values.
- Capture your early ideas on goals, OKRs and KPIs. These don't have to be perfect; you'll revisit them later.
- Create a materiality analysis matrix for your sustainable innovation programme and projects. Consider potential themes to include and stakeholders to invite input from.
- Reflect on how you might move towards or strengthen use of a Whole Life Value (WLV) approach to sustainable innovation in your organization.
- Identify skills and mindset strengths and development areas for you and your team if you have one.

Notes

1. O'Brien, D, Main, A, Kounkel, S and Stephen, A. Purpose is everything, Deloitte Insights, 15 October 2019, www.rand.org/pubs/reports/R4246.html (archived at https://perma.cc/V6EK-N5LN)
2. Ibid.
3. Skanska (2023) Our purpose and values, www.skanska.co.uk/about-skanska/skanska-in-the-uk/purpose-and-values/ (archived at https://perma.cc/QS4K-49DR)
4. Ferrovial (n.d.) Purpose and values, www.ferrovial.com/en-gb/company/purpose-vision-and-values/#:~:text=Our%20Purpose&text=We%20want%20to%20contribute%20to,for%20a%20world%20in%20motion (archived at https://perma.cc/DMM6-BPEV)
5. Maersk (n.d) About, www.maersk.com/about (archived at https://perma.cc/E4YK-M32A)
6. National Nuclear Laboratory (n.d.) Our purpose and values, www.nnl.co.uk/about/our-vision-and-values/ (archived at https://perma.cc/4SGA-6C42)
7. United Nations Global Compact (n.d.) Charting the course to sustainability, unglobalcompact.org/take-action/leadership/integrate-sustainability/roadmap (archived at https://perma.cc/4HJT-VL7K)
8. United Nations Global Compact (n.d.) Sustainability Stages Model, https://d306pr3pise04h.cloudfront.net/docs/issues_doc%2Flead%2Froadmap%2Froadmap-five-sustainability-stages.pdf (archived at https://perma.cc/VDK3-B2K6)
9. Plungis, J. Volkswagen emissions scandal: Forty years of greenwashing – the well-travelled road taken by VW, *The Independent*, 25 September 2015, www.independent.co.uk/news/business/analysis-and-features/volkswagen-emissions-scandal-forty-years-of-greenwashing-the-welltravelled-road-taken-by-vw-10516209.html (archived at https://perma.cc/ZD37-2P83)
10. Hotten, R. Volkswagen: The scandal explained, BBC, 10 December 2015, www.bbc.co.uk/news/business-34324772 (archived at https://perma.cc/5XQ2-4ZP3)
11. Kell, G. From emissions cheater to climate leader: VW's journey from dieselgate to embracing E-Mobility, Forbes, 5 December 2022, www.forbes.com/sites/georgkell/2022/12/05/from-emissions-cheater-to-climate-leader-vws-journey-from-dieselgate-to-embracing-e-mobility/ (archived at https://perma.cc/3JAB-4RCX)
12. Deloitte (2023) Deloitte Study: while most business leaders believe the right workplace model is key to success, only 24% are very ready to make needed changes, www2.deloitte.com/hr/en/pages/press/articles/Global-Human-Capital-Trends-2023.html (archived at https://perma.cc/4VSM-8LQT)

13 Scott, M. How a one-man scrap metal recycler became the world's most sustainable corporation, Corporate Knights, 18 January 2023, www.corporateknights.com/rankings/global-100-rankings/2023-global-100-rankings/top-company-profile-schnitzer-steel/ (archived at https://perma.cc/5ZGN-MVUG)

14 Radius Cycling (2023) Schnitzer Steel rebrands as Radius Recycling to reflect company's vision, purpose, and impact on circular economy, www.businesswire.com/news/home/20230726813790/en/Schnitzer-Steel-Rebrands-as-Radius-Recycling-to-Reflect-Company%E2%80%99s-Vision-Purpose-and-Impact-on-Circular-Economy (archived at https://perma.cc/YG2G-NQWX)

15 Asian Infrastructure Investment Bank (n.d.) Who we are, www.aiib.org/en/about-aiib/index.html (archived at https://perma.cc/YM5Z-J2A7)

16 Asian Infrastructure Investment Bank (n.d.) Infrastructure for Tomorrow, www.aiib.org/en/about-aiib/who-we-are/infrastructure-for-tomorrow/overview/index.html (archived at https://perma.cc/6DKF-7J4D)

17 Maersk (n.d.) Vision, www.maersk.com/about/vision (archived at https://perma.cc/C77Z-G7XR)

18 Government of Dubai (n.d.) Explore RTA, www.rta.ae/wps/portal/rta/ae/home/about-rta/explore-rta#abtMissionVision (archived at https://perma.cc/3WAT-9BX8)

19 Clean Path NY (n.d.) www.cleanpathny.com (archived at https://perma.cc/N5QS-GQXT)

20 Government of Dubai (n.d.) Explore RTA, www.rta.ae/wps/portal/rta/ae/home/about-rta/explore-rta#abtMissionVision (archived at https://perma.cc/9JRH-2P7A)

21 Clean Path NY (n.d.) www.cleanpathny.com (archived at https://perma.cc/N5QS-GQXT)

22 Office of Future Transport Technology (n.d.) Our vision, www.infrastructure.gov.au/infrastructure-transport-vehicles/transport-strategy-policy/office-future-transport-technology/our-vision (archived at https://perma.cc/N5QS-GQXT)

23 North, J (n.d.) Facilitation ideas for mission, vision & values, *The Big Bang Partnership*, https://bigbangpartnership.co.uk/facilitation-ideas-for-mission-vision-values/ (archived at https://perma.cc/Q3HP-GQRD)

24 Maersk (n.d.) Maersk core values, www.maersk.com/about/core-values (archived at https://perma.cc/K2TM-W9S7)

25 Government of Dubai (n.d.) Explore RTA, www.rta.ae/wps/portal/rta/ae/home/about-rta/explore-rta#abtMissionVision (archived at https://perma.cc/T6M7-H628)

26 Microsoft (n.d.) What we value, www.microsoft.com/en-pk/about/values (archived at https://perma.cc/34U3-HAX4)

27 As part of its mission, President of Microsoft Brad Smith, Chief Financial Officer Amy Hood, and CEO Satya Nadella, announced Microsoft's environmental commitments in 2020. These commitments are ambitious goals, which Microsoft called 'moonshots'.

28 Smith, B. Microsoft will be carbon negative by 2030, Official Microsoft Blog, 16 January 2020, blogs.microsoft.com/blog/2020/01/16/microsoft-will-be-carbon-negative-by-2030/ (archived at https://perma.cc/34U3-HAX4)

29 Jorgensen, M. How to turn your sustainability into action, Microsoft Blogger Series, 6 January 2022, www.microsoft.com/en-gb/industry/blog/cross-industry/2022/06/01/how-to-turn-your-sustainability-ambition-into-action (archived at https://perma.cc/2HNQ-FWU2)

30 Microsoft Corporation (2015) Datacenter Sustainability Strategy Brief, download.microsoft.com/download/1/1/9/119CD765-0CEE-4DA6-B396-20603D3F4701/Datacenter_Sustainability_Strategy_Brief.pdf (archived at https://perma.cc/7GRB-55UM)

31 Microsoft Corporation (2022) 2022 Microsoft Sustainability Report, query.prod.cms.rt.microsoft.com/cms/api/am/binary/RW15mgm (archived at perma.cc/69EP-LDDF)

32 Cherise. Microsoft's Commitment to Sustainability, Foundation IT, 5 September 2023, www.foundation-it.com/news/microsofts-commitment-to-sustainability (archived at https://perma.cc/CUD8-HK9L)

33 Microsoft (n.d) Powering sustainable transformation, datacenters.microsoft.com/globe/powering-sustainable-transformation (archived at https://perma.cc/6SFJ-ZE2F)

34 Cherise. Microsoft's Commitment to Sustainability, Foundation IT, 5 September 2023, www.foundation-it.com/news/microsofts-commitment-to-sustainability (archived at https://perma.cc/3ZD9-T92H)

35 Northumbrian Water Group (n.d.) About Northumbrian Water Group, www.nwg.co.uk/about-us/nwgroup/nwg-structure/ (archived at https://perma.cc/D2AC-3JPB)

36 Northumbrian Water Group (n.d.) Our vision and values, www.nwg.co.uk/about-us/nwl/what-we-do/our-vision-and-values/ (archived at https://perma.cc/8KPJ-BB92)

37 Northumbrian Water Group (2024) Northumbrian Water builds for AMP8 and beyond with largest-ever contractual commitment, www.nwg.co.uk/news-and-media/news-releases/northumbrian-water-builds-for-amp8-and-beyond-with-largest-ever-contractual-commitment/ (archived at https://perma.cc/YLF8-ER7K)

38 Siemens x Monocle. (n.d.) Wunsiedel: How a community is shaping their sustainable future, Monocle, monocle.com/content/Wunsiedel-How-a-community-is-shaping-their-sustainable-future/ (archived at https://perma.cc/Z8GV-GBJL)

39 Roads and Transport Authority (2021) RTA Sustainability Report 2021, www.rta.ae/wps/wcm/connect/rta/0bc9d171-4cf9-4cae-be22-3c1990180522/RTA-Sustainability-Report-2021-eng.pdf?MOD=AJPERES&CACHEID=ROOTWORKSPACE.Z18_N004G041LOBR60AUHP2NT32000-0bc9d171-4cf9-4cae-be22-3c1990180522-ohgA88i (archived at https://perma.cc/JE62-HDK9)

40 Roads and Transport Authority (2022) RTA Sustainability Report 2022, www.rta.ae/wps/portal/rta/ae/home/about-rta/sustainability (archived at https://perma.cc/F3BJ-SD9U)

41 Husqvarna Group (2022) Sustainovate Report 2022, www.husqvarnagroup.com/sites/husqvarna/files/2023-03/Husqvarna_ Sustainovate_Report_2022.pdf (archived at https://perma.cc/MC2K-ZFFG)

03

Mapping the innovation landscape: Horizon scanning

CHAPTER OVERVIEW

This chapter addresses the dynamic nature of the innovation landscape, focusing on rapid new technology development, societal and regulatory changes, economic pressures and the introduction of early-stage innovations to the market. It provides a guide for effective horizon scanning.

Contents and Sustainable Innovation Roadmap resources include:

- Understanding the dynamic nature of the innovation landscape, drivers and trends in sustainable innovation.
- How to engage in effective horizon scanning.
- Steps to prioritize potential future outcomes identified in the horizon scanning process, based on degrees of importance and certainty.
- The importance of stakeholder engagement.
- Tools to create scenario planning summaries that can be tested with key stakeholders and experts using methods such as interviews and the Delphi technique, in conjunction with additional data.
- How to build a transformation map for sustainable innovation, by understanding the current position, and 'backcasting' the likely steps to achieve your sustainable innovation projects.

The innovation landscape

Achieving net zero, advancing digital technology and shifting social values and priorities are key drivers of change at this point in the 21st century. As a result, the Volatility, Uncertainty, Complexity and Ambiguity (VUCA)[1] that organizations face today are unprecedented. Leaders of sustainable innovation need to navigate unchartered, shifting territory that can, on the face of it, make planning seem pointless. Why bother planning when we know things are going to change? Well, it is *because* there is so much VUCA that planning is necessary to steer the best course through to achieving the results we are aiming for. And an essential element of that planning process is building in dynamic capability, which is the ability to flex and adapt quickly in response to changes in the external environment.

Successful innovators are always on the lookout for the early signs of these changes (known as weak signals, and patterns) in the external environment. These signals are often subtle, easy to overlook and may not have immediate or obvious implications. They can come from various sources, including technological advancements, shifts in consumer behaviour, new regulations or changes in market dynamics. Leaders of sustainable innovation need to anticipate and be ready for change, rather than being caught out by it. They spot opportunities in these changes, as well as threats or challenges, and are proactive in taking steps to integrate the changes into their planning and risk management strategies. Sitting back and waiting to see what happens is not an option. Organizations that do so find themselves falling further behind, until the time comes when they are forced to take significant action to avoid becoming irrelevant. The world is always evolving, and it's unlikely that we'll reach a point where all uncertainties are resolved and everything becomes clear. We must engage with change and act, otherwise we won't get any feedback or information on the effectiveness of our ideas and strategies.

THE COVID-19 PANDEMIC: VUCA IN ACTION

The Covid-19 pandemic is a prime example of a VUCA environment:

Volatility

Volatility means instabilities of an unknown duration. The rapid spread of the virus and its mutations, alongside fluctuating infection rates across different regions, created highly volatile conditions. Businesses and individuals had to adapt quickly to changing restrictions and economic conditions.

Uncertainty

There was significant uncertainty about how the disease spread, its impacts and the duration of the pandemic. This uncertainty affected decision-making processes at all levels, from governments to individuals.

Complexity

Complexity means having many interconnections, interdependencies and variables. The complexity of the pandemic was evident in the interconnectedness of global supply chains, healthcare systems and the international travel network. Responses had to consider a myriad of factors, including economic impacts, public health and social behaviour.

Ambiguity

Ambiguity is present when brand-new situations occur and previous experience is of limited use in decision-making. Ambiguity in the pandemic was present in the varied symptoms of the virus, the effectiveness of different interventions and mixed messages from authorities. This led to confusion and differing opinions on how best to respond to the crisis.

The VUCA nature of the pandemic required responses that were flexible, adaptive and resilient, as traditional models of prediction and planning were often inadequate.

IMPACT OF VUCA AT PORT OF TYNE

The River Tyne in North East England has hosted a port for more than 2,000 years, playing a pivotal role for communities from the Roman era to the present day.

The Romans used the port to trade grain, wood, hides, salt, wool and fish in exchange for wine, leather, cloth, tiles and metal from Northern Europe, Spain and Italy.

Trade continued through the Middle Ages, followed by a period of newfound prosperity in the 1600s through exporting coal. Trade in coal was further fuelled by the Industrial Revolution and continued to grow,[2] until the shutdown of many mines in the region and the global decline in coal usage as a fuel hit the Port of Tyne hard. In 2016, the Port reported a 22 per cent drop in turnover versus the previous year, due to what it described as a 'dramatic reduction in coal volumes'.[3] The last ever shipment of coal left the Port of Tyne in 2021.[4]

In 2016, the Port of Tyne faced a critical situation due to the changing business environment and shifting market demands.

Volatility – shifts in the market

Due to the mine closures and the move away from coal-fired power stations for decarbonization, the port's income, once 70 per cent reliant on coal, vanished in only a few years.

Uncertainty

There was significant uncertainty about emerging new income sources to replace the income from coal. The Port of Tyne invested £14.6 million in infrastructure development, with the majority, £8.8 million, allocated for facilities to manage wood pellets imported for the Lynemouth Power Station in Northumberland, which was being refitted for renewable energy production. However, beyond those decisions, how the renewables market might evolve and potential sources of income to replace coal were uncertain.

Complexity

The complexity of the wider decline in coal was evident across the region. It affected local jobs, businesses, infrastructure and investments across the north-east region as a whole. The Port's long-standing customer relationships either transformed significantly or disappeared completely.

Ambiguity

The Port's deep expertise in managing coal shipments had limited transferability when it came to having to navigate an emerging renewable energy landscape.

> As a result of the dramatic impact of the decline of the coal industry on the Port of Tyne, the management team at the time faced unprecedented VUCA challenges that would take several years to recover from. With a new management team in place, the Port's successful turnaround began. We will discover how the Port of Tyne created its sustainable value proposition, innovated its business model and managed its legacy assets from Mark Stoner, the Port's chief financial officer, in Chapters 4 and 5.

How to plan amidst the realities of VUCA

Tuning into the environment and building the capability to be flexible, agile and collaborative are key skills for successful, adaptable planning within a VUCA environment:

Volatility: Build in an additional, affordable resource buffer that you can flex up or down quickly. How much time, investment and focus you give this needs to reflect the scale of the risk to your organization.

Uncertainty: Follow developments closely. Collect and use market and other intelligence in your decision-making. Create options and 'what if' scenarios.

Complexity: Work with people and organizations who bring subject matter expertise. Form collaborative, cross-functional teams to shape more holistic, integrated solutions.

Ambiguity: Adopt lean innovation principles to test innovations on a small scale, learn and retest on an iterative basis to get data and insight, make progress and reduce risk.[5]

How to plan for sustainable innovation amidst the realities of VUCA is a key theme throughout the following chapters. Your Sustainable Innovation Roadmap resources and activities will help you to lead your organization to become future-ready.

The Three Horizons Framework

It's helpful to think of any period of significant transition that brings VUCA via the lenses of the Three Horizons Framework.[6] Each horizon represents a different domain of innovation and organizational focus.

Horizon 1: maintain and improve core operations. Horizon 1 is about optimizing and extending the current business model and core capabilities. It's the present day, where an organization focuses on improving performance to deliver on short-term objectives.

Horizon 2: develop emerging opportunities. This horizon focuses on transitioning from the current state to future possibilities. It's where the seeds for future growth are sown, involving ventures that hold promise for the future but are currently unproven.

Horizon 3: create completely new business. The third horizon represents the future. It involves creating new opportunities that do not yet exist. This is the space for radical innovation, exploring future trends and developing entirely new business models or products that will define future markets.

Leaders of sustainable innovation need a balanced portfolio approach to strategic planning, ensuring that their programmes and projects are not solely focused on the present but are also actively developing future opportunities and innovations. There is usually a tension between the three horizons. Wherever possible, short-term decision-making needs to be future-focused to accelerate readiness and 'future-proofing' for Horizon 3.

ALIGNING SHORT-TERM DECISION-MAKING WITH READINESS FOR HORIZON 3

Here are the 10 key steps to aligning your short-term decision-making with future readiness and 'future-proofing' for Horizon 3:

1. Define Horizon 3 goals: Clearly outline what Horizon 3 entails for your organization. Understand the long-term objectives and the anticipated changes in your industry.
2. Assess current capabilities: Evaluate what resources, skills and processes you currently have and how they match up with the needs of Horizon 3.

3. Identify gaps: Determine what capabilities you lack to achieve your Horizon 3 goals.
4. Develop a strategy: Formulate a strategy that incorporates short-term actions with long-term objectives. This should include innovation, technology adoption and skills development.
5. Implement lean, flexible practices: Adopt lean, flexible practices that allow for quick response to change while keeping long-term goals in view.
6. Invest in learning and development: Enhance the skills and knowledge of your team to prepare for future demands. Continuously look outwards to spot early signs of impending change.
7. Measure and adapt: Establish key performance indicators (KPIs) that align with Horizon 3 objectives. Regularly review and adjust your approach based on these metrics.
8. Cultivate a forward-thinking culture: Create an organizational culture that values long-term thinking and continuous improvement.
9. Leverage data and trends: Use data analysis and trend forecasting to inform your decisions and anticipate future challenges and opportunities.
10. Review and refine regularly: Continuously revisit your strategy and operations to ensure they are in sync with evolving Horizon 3 scenarios.

The rest of this chapter, and the chapters that follow, will help you put each of these steps into practice.

Horizon scanning

Horizon scanning is the systematic process of gathering information to identify potential future risks, opportunities and developments for

Horizons 2 and 3. It is an essential activity for leaders of sustainable innovation because it helps with:

- **Anticipating change:** Horizon scanning focuses on spotting upcoming trends, allowing innovation leaders to be proactive rather than reactive.
- **Risk management and resilience:** By spotting potential future challenges, organizations can develop strategies to mitigate them.
- **Opportunity recognition:** Identifying emerging new customer needs, technologies or practices can give organizations a competitive edge and accelerate sustainable solutions.
- **Stakeholder engagement:** Understanding future landscapes ensures businesses remain aligned with stakeholder expectations.
- **Resource allocation:** Leaders can allocate resources effectively by knowing which sustainable innovations will have the most impact.
- **Regulatory compliance:** Staying ahead of potential future regulations ensures that organizations remain compliant and can prepare for any changes well in advance, avoiding a significant peak in effort and expense.

Horizon scanning should be an ongoing, daily process, with deeper dives at least once or twice a year to take stock and reset. Some organizations resist horizon scanning, arguing that it diverts time and resources from immediate concerns, potentially at the expense of current performance. These organizations tend to be focused on Horizon 1 only, and risk needing to operate in a highly reactive state much of the time. There is also the view that its usefulness, reliability and accuracy are dubious, especially in fast-changing markets. Plus, horizon scanning could lead to an overemphasis on new, unproven areas at the expense of core business activities. All that said, the risks of complacency, denial or even fear of change, and failure to look forward and outward, are much greater than any of the concerns about horizon scanning. This is as long as all three horizons – the now, transition and future state – continue to be well-balanced.

Foresight, forecasting and prediction

Horizon scanning is about foresight and forecasting, not prediction. While all three aim to understand the future, they are not the same (see Figure 3.1). Foresight is broad and exploratory. Forecasting uses data to estimate outcomes, and prediction asserts a specific future event. For example, the following statement is a well-informed, reliable forecast, based on foresight:

> If the world is to have a credible chance at limiting global warming to 1.5°C to avoid the worst impacts of climate change, global carbon dioxide (CO_2) emissions need to reach net zero by the early 2050s.[7]

Even though the example is based on substantial data and detailed, expert modelling, it is hypothetical (if x, then y...), no specific date is mentioned, and the statement is not predicting a specific outcome.

FIGURE 3.1 Comparison of foresight, forecasts and prediction

Feature	Foresight	Forecasts	Prediction
Purpose	Explores multiple future possibilities	Estimate a specific future outcome	States a specific future event
Method	Qualitative and quantitative	Mainly quantitative	Quantitative or qualitative
Time horizon	Long term	Short to medium term	Can be short or long term
Certainty	Low; considers various scenarios	Moderate; based on data trends	High; specific outcome expected
Usage	Strategy development, innovation	Planning, budgeting	Decision-making
Flexibility	Adaptable to change	Periodic updates required	Often fixed

There is no such thing as perfect information, especially about the future. The further ahead we look, the less reliable information becomes. This can be challenging for leaders of sustainable innovation who are creating plans for new programmes and projects that could last for a decade or so and often much longer. The intention of horizon scanning is to strengthen foresight so that we can forecast more accurately, always bearing in mind that our forecasts are informed estimates rather than definite predictions. It's about gathering intelligence about the future, exploring the dynamics of change and developing, and testing potential scenarios and strategies.

A process for horizon scanning

PESTEL ANALYSIS

For effective horizon scanning, it's important to gain insights from multiple and diverse sources and methods, and then combine them to identify potentially significant patterns and trends. A good place to start is with a PESTEL analysis, which has become a mainstay in strategic innovation analysis for many organizations. PESTEL analysis is an extension of ETPS,[8] or PEST analysis, which is an acronym for Political, Economic, Social and Technological factors used in the environmental scanning component of strategic management. Over time, to account for the increasingly complex business environment, two additional factors were included. These are Legal and Environmental, considering the impact of regulation and ecological aspects, respectively.

PESTEL (Figure 3.2) provides a useful checklist of areas to consider when horizon scanning. Ideally, your team, armed with data and research, will work through each factor, capturing the emerging developments for each and highlighting the potential impacts in the context of your sustainable innovation strategy, programme or project. It's critically important not to stop there. PESTEL warrants more than a brainstorming session, which is merely the jumping-off point. Potentially useful or important threads need to be explored further, informed by still more data, ideally using some of the methods explored later in this chapter.

FIGURE 3.2 Sustainable Innovation Roadmap: PESTEL canvas

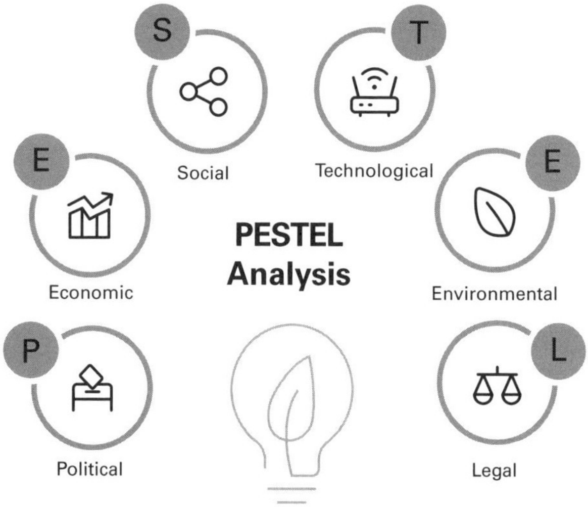

Some PESTEL themes for leaders of sustainable innovation to consider are detailed below.

POLITICAL FACTORS

As discussed in Chapter 1, in 2015 the United Nations set out its 17 ambitious Sustainable Development Goals aiming to guide global efforts towards sustainability. However, the real-world impact of these goals is still up for debate. Biermann et al (2022) sifted through an enormous pile of research – more than 3,000 studies from 2016 to April 2021 – to see what influence, if any, the UNSDGs have had on political actions and policies.[9] They found that these goals are indeed making a difference. They've influenced sustainability conversations from local councils to international summits. Yet, when it comes to actual, tangible changes – like enacting new laws or shifting financial priorities to support these objectives – there is much more to be done. The UNSDGs haven't yet reshaped the political landscape as much as might have been hoped, but some progress is being made.

Political factors such as taxation and duty, grant funding, international trade policy, environmental policy, priorities and stability will vary from region to region. Geopolitical factors can have wide-reaching impacts across regions. Geopolitical risks can significantly impact global supply chains. A 2023 study by Qin et al illustrates this.[10] Infrastructure damage during conflicts, like the Kosovo war in 1999, for example, obstructed supply lines. Wars in Afghanistan, Iraq and Syria also illustrate such disruptions. Sanctions, as seen in the Russia–Ukraine conflict, further strain global trade by limiting resource exchange and increasing trade barriers. For instance, sanctions on Russia affected sectors like energy and technology. Oil production fluctuations during geopolitical tensions, as in the 2003 Iraq war, can spike oil prices, increasing transportation costs and pressuring supply chains. However, this isn't always the case; during the 2014 Russia–Ukraine conflict, oil prices fell, mitigating supply chain threats. There is a complex relationship between geopolitical events and global supply chain stability, an important consideration for leaders of sustainable innovation.

ECONOMIC FACTORS

For sustainable innovation in technical, infrastructure and other major projects, there are many economic factors to consider as part of a PESTEL analysis, such as:

- **Interest rates:** The cost of borrowing money to fund projects. For example, if a company wants to build a new green energy facility, the interest rates will affect the overall cost of the project.
- **Economic growth:** The general health of the economy can influence infrastructure development. In a strong economy, there may be more investment in sustainable projects. For example, the European Union's strategy for Green Infrastructure focuses on protecting, rehabilitating and improving natural environments to halt the decline of wildlife diversity and ensure that ecosystems continue to benefit society. Additionally, the EU's 2030 plan for biodiversity emphasizes the need for funding in natural and aquatic habitats and the consistent inclusion of robust ecosystems, green spaces and solutions inspired by nature within city development frameworks.[11]

- **Inflation rates:** Higher inflation can increase the cost of raw materials, affecting budgets for constructing new technology or infrastructure.
- **Exchange rates:** For projects that require importing materials or technologies, the strength of the currency can impact costs significantly.
- **Government funding:** Grants, subsidies or tax incentives for sustainable projects can be a major economic factor, making some projects feasible that wouldn't be otherwise. For example, the United States federal tax credits for renewable energy,[12] which provide financial incentives for the development and installation of solar and wind power projects.
- **Market demand:** The level of demand for sustainable innovations affects investment and development. For instance, increased demand for electric vehicles spurs infrastructure development for charging stations. By the close of 2022, the world saw a rise to 2.7 million public EV chargers, with more than 900,000 new ones coming online in just that year. This marked an increase of 55 per cent from the previous year and matched the 50 per cent annual expansion seen from 2015 to 2019, before the global pandemic.[13]
- **Resource costs:** The price of commodities like steel or concrete can directly affect the financial viability of major projects.
- **Labour costs:** The availability and cost of skilled labour necessary to implement sustainable technologies is another economic factor.
- **Economic policies:** Policies that affect industry sectors relevant to sustainability, such as renewable energy targets or carbon pricing.

PESTEL ECONOMIC FACTORS: AN EXAMPLE OF THE IMPACT OF ECONOMIC POLICIES ON SUSTAINABLE INNOVATION

US adjustments to trade tariffs are an example of how economic policies in PESTEL can impact sustainable innovation.

In 2022, in the US, President Biden decided to adjust trade policies on solar products that were set up by the previous administration.[14] Some tariffs on imported solar technology remained, with a tax of 14–15 per cent on solar panels made from crystalline silicon (which is the most common type used in solar energy systems). However, President Biden made some changes to allow more solar products to enter the US without being taxed. These changes included doubling the quantity of solar cells that could be imported tariff-free and starting discussions with Canada and Mexico to let them send their solar goods to the US without any tariffs. As a continuation of existing policy, tariffs would not be applied to bifacial panels, however. Bifacial panels can absorb sunlight on both sides.

President Biden's move was intended to help America get ahead when switching to clean energy. These changes were trying to strike a balance. They aimed to keep American solar manufacturers in business by protecting them from a flood of cheap imports, mainly from China. But, at the same time, it was important to make sure that solar energy projects in the US could keep growing via access to affordable solar products from abroad.

Some domestic solar manufacturers weren't happy, though, because they felt this move didn't protect them enough and could hurt American jobs and investments. They were concerned it would give China an advantage in the solar industry. On the other hand, many companies that install solar systems thought that this was a good decision because it meant they could still use less expensive imported panels, helping to keep costs down.

- **Costs and benefits of innovations:** Sustainable innovations often require a significant upfront investment, which can be a barrier to implementation. However, the long-term benefits, such as reduced environmental impact, energy savings and improved social goodwill, often outweigh initial costs. For instance, incorporating smart grid technology into infrastructure projects has a high initial cost but leads to long-term savings through efficient energy use and decreased waste.
- **Investors and finance:** The role of investors is crucial in driving sustainable innovation. Investment decisions are increasingly influenced by the potential for sustainable growth and long-term value creation that sustainable innovations create.

PESTEL ECONOMIC FACTORS: EXAMPLES OF TYPES OF FINANCE FOR SUSTAINABLE INNOVATION PROJECTS

The availability of external financing of sustainable innovation can be an essential economic factor in PESTEL that influences whether or not many major projects go ahead.

- **Sustainable finance**: This refers to financial services integrating environmental, social and governance (ESG) criteria into business or investment decisions.[15] It's a trend growing in importance, with more funds flowing into projects that support sustainable energy and infrastructure development.
- **Blue finance**: Blue finance supports projects that protect and sustainably use ocean resources.[16] This can mean funding for the development of marine renewable energy or for infrastructure that improves the health of marine ecosystems.
- **Conservation finance**: Conservation finance focuses on investing in projects that aim to protect biodiversity and maintain ecosystems.[17] For infrastructure, this could involve funding the integration of green spaces into urban design, thereby improving air quality, and providing community health benefits.
- **Blue economy**: Like blue finance, the blue economy refers to economic activities that use the sea and its resources for economic growth, improved livelihoods and jobs while preserving the health of ocean ecosystems.[18] Sustainable port infrastructure and reduced-pollution shipping routes are examples of the blue economy in action.
- **Green economy**: A green economy prioritizes sustainability, reducing environmental risks and ecological scarcities. It is also socially inclusive.[19] In infrastructure and technical projects, examples are investments in renewable energy plants or the use of sustainable materials in construction that not only serve to reduce carbon footprints but also generate economic growth through job creation.

Economic factors in PESTEL analysis demonstrate the need to balance short-term financial outlays with the long-term economic, environmental and social returns on investment. The relationship between innovative technologies and new financial solutions is key to funding the transition towards more sustainable infrastructure and technical innovation projects.

SOCIAL FACTORS

Social factors in PESTEL are the cultural and demographic aspects of the external environment that can affect an organization. Examples of relevant social themes for sustainable innovation in technical, infrastructure and other major projects are:

- **Demographics:** Population size, age distribution, urban versus rural distribution and family size can influence the demand for and design of infrastructure projects.
- **Cultural trends:** Public interest in sustainability, attitudes towards technology and innovation and preferences for certain types of infrastructure (e.g. green spaces in urban areas) can shape project outcomes.
- **Education levels:** The education and skill level of the local population may determine the complexity of technologies that can be successfully implemented and maintained.
- **Health consciousness:** A society's focus on health might influence projects to prioritize features that promote physical activity, like bike lanes, or that reduce pollution.
- **Lifestyle changes:** Shifts in work-life balance, remote working trends and leisure activities could alter transportation needs and energy consumption patterns.
- **Social mobility:** The ease with which individuals can change their socioeconomic status may influence project design to improve inclusivity and accessibility.
- **Community engagement:** The degree to which the local population is involved in decision-making processes for new projects can impact their success and acceptance.
- **Consumer behaviour and expectations:** Preferences for sustainable products and services can drive innovation towards more environmentally friendly solutions.
- **Income distribution:** Wealth distribution in a region can influence the type of projects that are viable, as well as their scale and accessibility.

- **Ethical priorities:** The growing requirement by society for ethical behaviour from companies is a significant factor. The expectation is that organizations will go above and beyond legal compliance and treat employees, customers, suppliers and stakeholders with fairness and integrity. This means that having clear internal standards and policies for matters such as anti-corruption, diversity, inclusion and equity, demonstrating commitment to promises by taking action, responsible management of personal data, and more, are factors that are critical to brand reputation.

TECHNOLOGY FACTORS

Technological developments offer an unprecedented opportunity to transform and accelerate sustainable innovation in technical projects. There are several developments to consider, including:

Artificial intelligence (AI)

During 2022 and 2023, AI has seen unprecedented growth and the fastest adoption of new business technology in history.[20]

Applied AI is designed to perform a specific task or set of tasks. It operates within a predefined framework or ruleset to carry out functions like image recognition, language translation or routine customer service enquiries. These systems execute programmed responses based on the data they receive.

Generative AI, on the other hand, refers to systems that can generate new content or data that is not from their training set. This type of AI uses techniques like machine learning to analyse and learn from input data, and then works to create new, original output that can include text, images and code. Examples of generative AI include chatbots that can generate human-like text and AI that can compose music or create realistic-looking images or videos from scratch.

While there is much to learn and resolve about the legal, ethical and safety concerns of AI, for examples issues such as copyright, cybersecurity and user confidence in the accuracy and appropriateness of AI outputs, it's a technology development that seems to have high growth potential for sustainable innovation use cases. McKinsey's Technology Trends Outlook 2023[21] reported that:

- Between 2018 and 2022, applied AI topped innovation scores and stayed among the top five for investments. In 2022, it also saw the highest demand for specialists.
- Generative AI's low rankings in interest and funding in 2022 were contrasted by a significant rise in searches, suggesting growing curiosity and expected growth.

Artificial intelligence can be used to forecast maintenance needs, optimize energy use, mitigate climate-related risks and much more.

Digital transformation

The move towards digital infrastructures offers streamlined workflows and improved interconnectivity. It facilitates superior project management systems, developments such as smart cities, buildings and digital twins, which are virtual models designed to simulate a physical object, system or process. Digital twins can model the impact of decisions before they are taken and predict issues before they occur, leading to more resilient infrastructure.

Drones, cameras and satellite technology are all redefining precision in tracking, surveying and inspecting, providing cost-efficient and accurate data critical for the protection and maintenance of physical assets.

Web 3.0

Web 3.0 envisions an evolved internet underpinned by blockchain technology.[22] This is a collective system where data and platforms are maintained by user-operated networks rather than a few dominant entities, leveraging the inherent decentralization of blockchains. Blockchain technology could lead to heightened transparency and security in managing supply chain and transactional data, and smart contracts.

Web 3.0 could reshape sustainable innovation projects in the future with its decentralized nature, enabling smoother collaboration across various stakeholders and borderless project management.

Smart contracts and tokenization will streamline project execution and incentivize sustainable practices. The secure data management inherent in Web 3.0 ensures sensitive information is handled safely, helping to build more trust in international collaborations. Access to new funding avenues and enhanced transparency, particularly in supply chains, will assist in ensuring that resources are ethically sourced and that organizations remain accountable for their projects. Also, the protection of intellectual property under Web 3.0 could encourage more innovation, securing the benefits for creators and leading to significant advances.

The metaverse

Still in development is the metaverse, a collective virtual shared space, created by bringing together virtual reality (VR), augmented reality (AR) and the internet. It's a potential future version of the internet, where users can interact with a computer-generated environment and other users, accessed via VR headsets, AR glasses, smartphone apps or other devices.

The metaverse is modelled as a highly immersive, interactive and hyper-realistic space where people can work, play, socialize and participate in a wide range of experiences that go beyond what's possible in the physical world. It will continue to exist when you're not logged in, and you'll be able to take items from one environment to another.

As the metaverse evolves, it could support the transition to Industry 5.0, which encourages the use of new technologies not just to increase output but also to improve the quality of work life for people, and to protect the planet. The metaverse can offer spaces for virtual interaction that reduce the need for physical movement, aligning with sustainable consumption and production patterns. While the full impact of the metaverse on sustainability is yet to be seen, its potential to facilitate responsible digital transformation and support sustainable development goals looks promising.[23,24]

WORLD ECONOMIC FORUM'S 2023 LIST OF TOP 10 EMERGING TECHNOLOGIES

The World Economic Forum's 2023 list of top 10 emerging technologies includes flexible batteries, generative AI and sustainable aviation fuel, indicating pivotal shifts in energy storage, artificial intelligence and eco-friendly travel.[25] Designer phages and AI-facilitated healthcare are breakthroughs in medical treatment personalization, while the metaverse emerges as a tool for mental health. Wearable plant sensors and spatial omics highlight the deepening interface of biology and tech, leading to deeper environmental and health insights. The progression towards sustainable computing and flexible neural electronics points to a future where efficiency and human-centric design will combine.

ENVIRONMENTAL FACTORS

The energy sector is going through a transformative shift as it moves towards alternative sources. This transition demands systemic changes, such as constructing new production facilities and distribution networks. The potential for hydrogen as a clean fuel is leading to the creation of specialized hubs and the retrofitting of existing systems for its use and transportation. The shipping industry, led by players such as Maersk, is venturing into green methanol to power its vessels, with new infrastructure requirements for storage and refuelling.

PESTEL ENVIRONMENTAL FACTORS: TRANSFORMING SHIPPING – MAERSK'S JOURNEY TO NET ZERO

The carbon dilemma in shipping

Every year, ships burn 300 million tonnes of fuel to transport 11 billion tonnes of goods globally. This results in roughly 1,076 million tonnes of CO_2 emissions – about 3 per cent of the world's total greenhouse gas output.[26]

While new guidelines are helping to cut down emissions, efficiency alone won't meet Maersk's ambitious net zero target by 2040,[27] due to the fuel that's powering the world's ships.

The key to reaching net zero lies in changing what goes into the tank. Maersk has taken a bold step: to order new ships capable of running on green fuels.

Green fuels

Maersk's decision on which green fuels to use in their vessels is strategic and aimed at achieving net zero emissions by 2040. Several alternative fuels exist, but green methanol (e-methanol or bio-methanol) stands out for its feasibility and quick impact.

Maersk partnered with Ørsted, a leader in renewable energy, to develop a significant Power-to-X facility on the US Gulf Coast. This facility is designed to produce about 300,000 tonnes of e-methanol annually, which Maersk will use for its newly ordered fleet of 12 methanol-powered vessels.[28] The choice of green methanol as a fuel is based on its market readiness and scalability as a solution for shipping. The project represents a landmark effort in the green transformation of international deep-water shipping and sets a standard for future large-scale production of green maritime fuels.

Further collaboration includes securing green methanol supply solutions with other partners, like Proman, which aims to provide Maersk with 100,000–150,000 tonnes per year of green methanol from a new facility in North America. This facility plans to produce bio-methanol from non-recyclable forestry residues and municipal solid waste, with operations starting in 2025.[29]

Maersk teamed up with MAN Energy Solutions to retrofit dual-fuel methanol power into one of its container ships, a shipping industry first.[30] MAN Energy Solutions' MAN B&W ME-LGIM engine is a 10-metre-high powerhouse, with the output of 112 family cars.[31] When fuelled by green methanol, this engine brings Maersk much closer to carbon-neutral sailing. Maersk's 2023 vessel, powered by a dual-fuel engine, shows that green methanol is a viable option for the future. European Energy in Denmark produces e-methanol for Maersk's first green feeder vessel.[32]

The transition to green methanol and other sustainable practices requires a significant financial investment. Maersk has committed to investing billions over the next decade in new technologies and cleaner fuels.

Technology adoption

Maersk is not just relying on green methanol but is also exploring other technologies such as battery systems and hydrogen fuel cells for auxiliary power. These technologies can further reduce the overall carbon footprint of vessels.

Beyond the sea: a comprehensive approach

The company's decarbonization efforts extend to terminals, land transport and last-mile deliveries. For example, their electric mini delivery vehicles in Delhi mark a step towards a fully carbon-neutral supply chain.[33]

Collaborative progress

Maersk is not working on its decarbonization plans alone. Their partnerships span multiple sectors, for example:

EQUINOR

Based in Norway, Equinor is primarily engaged in the exploration, production and distribution of oil, gas and renewable energy. They are a critical player in energy resource development, particularly in sustainable energy.

HYUNDAI MIPO DOCKYARD

Located in South Korea, Hyundai Mipo Dockyard excels in the design, construction and repair of a wide variety of maritime vessels. They bring invaluable shipbuilding expertise to the table.

OCI GLOBAL

Headquartered in the Netherlands, OCI Global focuses on developing nitrogen, methanol and hydrogen-based solutions. Their work aims to reduce carbon emissions in sectors that are traditionally energy-intensive.

MAN ENERGY SOLUTIONS

A Germany-based company, MAN Energy Solutions offers a range of power generation technologies, including engines and turbines. They are leaders in creating innovative and sustainable energy solutions for various industries.

EUROPEAN ENERGY

Operating out of Denmark, European Energy is in the process of building a facility to produce green methanol. The facility aims to provide 16,000 tonnes of green methanol by 2024, directly supporting Maersk's decarbonization goals.

Regulation and policy

Maersk is actively participating in industry dialogues to advocate for tighter emission standards. The company is a member of various global shipping organizations that are working towards unified regulations for cleaner oceans.

Employee training

A crucial part of achieving net zero emissions is the workforce. Maersk is investing in training programmes to educate its employees on the importance of sustainability and how to operate new, greener technologies effectively.

CUSTOMER ENGAGEMENT

Maersk is introducing sustainability metrics and tracking for customers to view their own carbon footprint. This feature aims to make the supply chain more transparent and encourage customers to opt for greener shipping solutions.

COMPETITIVE ADVANTAGE

Maersk's early adoption of green technologies not only positions it as an industry leader in sustainability but also provides a competitive advantage. Customers looking to reduce their own emissions are more likely to partner with greener shipping options.

SCALABILITY AND FUTURE PLANNING

The partnership with European Energy for green methanol is just the beginning. Maersk is in talks with multiple other suppliers to ensure a larger scale of sustainable fuel options in the future.

SOURCE www.maersk.com/sustainability/all-the-way-to-net-zero

The growth of renewable energy sources and electric vehicles means a new era of infrastructure, requiring upgraded grid capacity and accessible, high volumes of charging stations. In parallel, nuclear energy and carbon capture technologies are also on the rise.

Renewable wind energy is growing fast, with the first phase of the world's largest windfarm at Dogger Bank, 130 km off the coast of North East England and spanning UK, German, Danish and Dutch waters, which became operational in 2023.[34] We'll explore Equinor's sustainable innovation activities at Dogger Bank in more detail in Chapter 10.

The potential of tidal energy is being developed as an affordable, consistent and sustainable power source,[35,36] though still in its early development stages.

Critical minerals, such as lithium, are essential for the advancement and sustainability of various clean energy technologies, especially for energy storage. The demand for batteries that can store energy efficiently – especially for electric vehicles and renewable energy systems – is growing fast. However, the extraction and processing of lithium and other critical minerals raise significant environmental and ethical concerns.[37] The mining of these minerals can lead to substantial ecological disruption, water use conflicts and other social challenges in local communities. Plus, the geopolitical landscape of mineral-rich countries often complicates the supply chain. This could lead to issues around the stability of supply and ethical considerations, such as labour conditions and fair-trade practices. Securing a sustainable and ethical supply chain for these critical minerals is essential.

There is a growing demand for leaders of sustainable innovation projects in sectors such as construction, utilities, agriculture, urban development, land management and infrastructure to integrate nature-based solutions (NbS) into innovation design and implementation. NbS are strategies that utilize natural processes and ecosystems to address environmental, social and economic challenges.

NbS solutions with safeguards are estimated to provide 37 per cent of climate change mitigation until 2030 needed to meet the goal of keeping climate warming below 2°C, according to the Intergovernmental Science-Policy Platform on Biodiversity and Ecosystem Services (IPBES).[38]

Just a few examples of NbS are:

- **Urban green spaces:** Parks, gardens and green rooftops that provide recreation, improve air quality and enhance biodiversity. For example, the High Line in New York City is a famous urban park built on a historic freight rail line elevated above the streets on Manhattan's West Side.
- **Green infrastructure for stormwater management:** Using vegetation and soils to manage rainwater where it falls, such as through rain

gardens, green streets and permeable pavements. The Sponge City Initiative in China is an example where cities are designed to absorb and reuse rainwater.[39]

- **Reforestation and afforestation:** Planting trees on deforested lands or on lands that have never been forested for carbon sequestration, biodiversity conservation and soil stabilization. The Great Green Wall across the Sahel region of Africa is a project aimed at reversing desertification. According to the UN Environment Programme,[40] 'by 2030, the Great Green Wall aims to restore 100 million hectares of land, sequester 250 million tonnes of carbon, and create 10 million jobs. It is providing food and water security, habitat for wild plants and animals, and a reason for residents to stay in a region beset by drought and poverty.'
- **Bioengineering for slope stabilization:** Using plants, especially deep-rooted species, to stabilize slopes and prevent landslides or erosion. An example includes the use of vetiver grass systems in various countries for erosion control.
- **Constructed wetlands for wastewater treatment:** Using engineered wetland systems to treat wastewater naturally through biological processes. For example, the East Kolkata Wetlands in India act as a natural sewage treatment plant.[41]

The International Union for Conservation of Nature (IUCN) Global Standard for Nature-based Solutions (2020) outlines guidelines for implementing NbS in projects.[42]

PESTEL ENVIRONMENTAL FACTORS: SUSTAINABLE INNOVATION IN ACTION AT THE OFFSHORE RENEWABLE ENERGY CATAPULT

The Offshore Renewable Energy Catapult (OREC) is the UK's leading technology innovation and research centre for offshore renewable energy. Its purpose is to 'play a key role in delivering the UK's net zero targets by accelerating the creation and growth of UK companies in the offshore renewable energy sector'. OREC uses its unique facilities and research and engineering capabilities to bring together industry and academia and drive innovation in renewable energy.[43]

OREC is focusing on four key areas for future innovation:

- Floating wind
- Marine energy
- Testing and demonstration
- Operations and maintenance

Each area has its own Centre of Excellence to drive innovation in robotics, autonomous systems, big data and artificial intelligence, balance of plant – especially foundations – and next-generation technologies in the sector.[44]

The rapid growth and importance of offshore renewable energy's contribution to decarbonization are evident in the pace and volume of sustainable innovation in the sector, demonstrated by OREC's activities. Between 2013, when it launched, and 2024, OREC achieved:

- £677 million total value of innovation projects.
- More than 650 research and development projects.
- Support for more than 1,350 small to medium-sized enterprises, including in OREC's technology accelerator programme for new companies pursuing offshore wind success.
- Publication of over 250 academic research papers.
- Commercialization of more than 148 new products and services.[45]

LEGAL FACTORS

A key factor in sustainable innovation is regulation. In some cases, such as in the construction industry in 2023,[46] regulation drives industries forward and accelerates progress. In others, for instance AI in the same year, regulation is behind the curve.[47]

Regulations and policies aimed at enhancing environmental, social and governance (ESG) standards are continually being developed and launched across the world. A central focus is on transparency, compelling companies to disclose sustainability practices and risks. The EU's Corporate Sustainability Reporting Directive (CSRD)[48] and the International Sustainability Standards Board (ISSB)[49] are at the forefront, pushing for global consistency in reporting standards.

The European Climate Law made the European Green Deal's ambition of a climate-neutral Europe by 2050 official. It mandates an interim reduction of net greenhouse gas emissions by at least 55 per cent by 2030 from 1990 levels, is legally binding and ensures that all EU actions align with the climate neutrality aim. The law introduced robust monitoring systems to track progress, informed by scientific insights and aligned with the Paris Agreement's global stocktakes to assess the impact of countries' climate actions. Additionally, the law emphasized the enhancement of the EU's carbon sink, with new regulations effective from May 2023.[50]

ESG considerations are transforming the investment world. Europe's Sustainable Finance Disclosure Regulation (SFDR)[51] is pioneering this shift, with the UK and countries in Asia Pacific following suit by introducing measures against greenwashing and for reliable ESG fund labelling.[52] In the US, policies like the Inflation Reduction Act are incentivizing the economic viability of emerging green technologies.[53]

Taxonomies, or consistent categorizations, are becoming a global tool for defining 'green' investments, with countries like China and the EU[54] leading the way in establishing standards that reflect their economic priorities. However, this brings risks of inconsistency across borders, highlighting the importance of efforts like the EU-China Common Ground Taxonomy[55] for alignment of international investments.

Globally, there's a movement towards mandatory transition plans for companies, demanding credible strategies to mitigate environmental and social harms and align with global climate goals. The UK's proposed broad framework for transition plans is an example of this, potentially shaping future mandatory requirements.[56] This could have several impacts. Companies will need to integrate sustainability deeply into their strategies, attracting investment by showcasing their commitment to environmental and social goals. Regulatory adherence will become essential, driving innovation and potentially giving rise to new market leaders in green sectors. Consumers, increasingly eco-conscious, may favour businesses with transparent and robust sustainability practices, influencing purchasing decisions. Operational shifts, such as adopting cleaner technologies and rethinking supply chains, will be critical. Organizations that lag in implementing credible transition plans risk legal consequences, financial penalties and a dent in their market reputation.

This legal momentum indicates a transition towards mandatory sustainability practices and detailed disclosure requirements, compelling companies, and investors, to align with these evolving legal standards to drive sustainable innovation.[57]

CASE STUDY

PESTEL factors in action at WUN H2

WUN H2 operates as a joint initiative between Siemens Financial Services, Rießner-Gase and Stadtwerke Wunsiedel to create one of Germany's top green hydrogen production facilities. The plant in Wunsiedel Energy Park began operations in September 2022. It harnesses wind and solar energy to produce up to 1,350 tonnes of green hydrogen each year, powered by an 8.75-megawatt electrolyser.[58] The project shows each PESTEL factor in action:

- Political: The Mayor's and Bavarian local government's support provided the political will to back Wunsiedel's green initiatives, essential for a project of this kind.
- Economic: Investment from companies such as Siemens Financial Services indicated the economic viability and interest in funding innovative sustainable energy solutions.
- Social: The project addressed social demands for cleaner energy and contributes to local job creation.
- Technological: Use of advanced electrolysers and integration with the existing energy park leveraged technological progress in renewable energy. A digital twin and collaboration with Siemens advanced the project's sustainable innovation.
- Environmental: By producing green hydrogen using renewable sources, the project directly contributed to decarbonization and showed the importance of environmental responsibility.
- Legal: Operating within Germany's stringent environmental regulations, the project set a precedent for compliance with green standards. It also demonstrated an innovative approach to ownership.[59]

These PESTEL factors have collectively influenced the project's implementation and success.

Beyond PESTEL

PESTEL analysis is instrumental in identifying opportunities and threats, aiding in strategic planning. Regular updates to this analysis are necessary to reflect the dynamic nature of these external factors.

It is crucial to involve your team and stakeholders in the process to ensure the best possible outcomes for your business.

Many organizations stop their horizon scanning once their PESTEL analysis is complete. In reality, it should just be the beginning. PESTEL gives you a useful checklist to work through, but your thinking will need greater depth if it is to be insightful and useful.[60]

Disruptive events

Alongside your work on PESTEL, it's also important to be aware that unexpected events can significantly disrupt normal operations or planned outcomes of sustainable innovation projects. While some disruptive events can be anticipated and planned for, like Brexit, others may arise unexpectedly, such as geopolitical conflicts or pandemics. Innovation resilience and contingency planning are needed, as unexpected events can hinder or halt progress. Of course, certain generally negative disruptions can unexpectedly accelerate change in some areas. The Covid-19 pandemic is an example of this, as it accelerated digital transformation.

Effective risk management strategies are needed to navigate sustainable innovation through unpredictable times, and Chapter 10 will delve into those risk management strategies in more detail.

MAKE LINKS ACROSS THE PESTEL FACTORS: CROSS-IMPACT ANALYSIS

Cross-impact analysis (Figure 3.3) is a technique that helps to identify the potential impact of combining different trends or variables from PESTEL analysis. Unlike the traditional approach of looking at individual categories, cross-impact analysis considers the interaction between two or more trends. For example, when the trend of an ageing population is combined with an increase in social media usage, the likely outcome of these two trends together can be considered. An outcome of this may be to consider digital inclusivity for the ageing population, for example. PESTEL factors don't exist in isolation, neatly separated from each other. One or more are usually inextricably linked, such as economic policy with politics. It's important to identify how different factors can influence each other to understand changes, rather than simply spot them. This is illustrated by the following quote from the US Energy Information Administration:

> Geographical, political, economic, and social circumstances will determine the exact pathways to transitioning energy systems in different parts of the world to net zero. However, broadly speaking, cutting carbon emissions from the global energy system requires accelerating the deployment of technologies to achieve two objectives: (i) electrifying as much of the energy demand as possible; and (ii) fully decarbonising electricity supply, using renewable sources.[61]

STAKEHOLDER ENGAGEMENT TO ACCESS DIVERSE PERSPECTIVES AND COMMITMENT

Interpretation of the insights extracted from a PESTEL analysis is likely to be prone to human biases and assumptions.[62] To prevent a limited view of potential future scenarios, it's essential to obtain data

FIGURE 3.3 Sustainable Innovation Roadmap: cross-impact analysis canvas

	Item A	**Item B**	**Item C**
Item A	A affects A	A affects B	A affects C
Item B	B affects A	B affects B	B affects C
Item C	C affects A	C affects B	C affects C

from diverse inputs and stakeholder perspectives, as recommended in the materiality analysis section of Chapter 2.

Stakeholder engagement refers to the process of involving individuals or groups who have a vested interest or influence in your organization's activities. Your stakeholders can include employees, customers, investors, suppliers, regulators, media and the local community, among others. Engaging with stakeholders goes beyond simply communicating with them; it involves actively seeking their input, involving them in decision-making processes and addressing their concerns and expectations.

Stakeholder engagement is not merely a nice-to-have; it's a best practice fundamental to sustainable development goals and corporate governance. The level of engagement with different people, from project team members to local communities, has a significant potential impact on the success of a proposed innovation project.[63]

Successful stakeholder engagement is vital for innovation success for several reasons. First and foremost, it enhances your organization's brand reputation. When stakeholders feel heard and valued, they become advocates for the organization, spreading positive word-of-mouth and attracting new customers. Moreover, engaged stakeholders are more likely to remain loyal to the brand, leading to increased customer retention.

Before your organization can effectively engage with its stakeholders, you first need to identify and understand who your stakeholders are and what their interests and concerns may be. Stakeholders can be categorized into internal and external stakeholders:

- Internal stakeholders include employees, board members and shareholders. They have a direct interest in the success of the organization and are directly affected by its decisions and actions.
- External stakeholders can include your entire sustainable innovation ecosystem – your customers, suppliers, investors, regulators and the local community. These stakeholders may have different expectations and concerns, and it's important to identify and address them accordingly. We'll go into much more detail on your sustainable innovation ecosystem in Chapter 10.

With multiple stakeholder groups, it can be challenging to identify and prioritize their needs and concerns. Conduct thorough stakeholder analysis and prioritize those who have the most influence and/or are impacted the most by your organization's activities and plans.

Tried and tested methods for stakeholder engagement in horizon scanning include the seven questions futures technique for scenario planning, the Delphi technique and the futures wheel. Use the insights from these stakeholder activities to inform your materiality analysis, discussed in Chapter 2.

Seven questions futures technique The seven questions futures technique (Figure 3.4) is a scenario planning approach that involves thinking about different possibilities for the future and working through each one. The approach can be used in different ways, such as through interviewing, questionnaires, surveys or focus groups. You can modify the questions to suit your specific industry or situational requirements.

I also recommend that, after a round of open questions, you use prompts that you have created from your own PESTEL analysis to access some more detailed responses on specific topics. Be sure to ask open-ended or neutral questions that allow your stakeholders to provide information without being directed towards a particular answer.

Here are the seven core questions:

1. If you could speak to someone from the future who had knowledge about your business area, what would you ask them?
2. What is your vision for success?
3. What are the dangers of not achieving your vision?
4. What changes need to occur in systems, relationships, decision-making processes and culture for your vision to be realized?
5. Looking back, what are the successes we can build on? The failures we can learn from?
6. What needs to be done now to ensure that your vision becomes a reality?

7 If you had absolute authority and could do anything, is there anything else you would do?

The seven questions technique helps you get a sense of where people see the future, what their hopes are and what they are worried or concerned about. It's a creative and engaging way of involving stakeholders in your horizon scanning. Capture all the outputs from the things people say, ready to analyse all your feedback in the round to spot trends and extract insights.

The Delphi technique The Delphi technique is a process of tapping into the collective wisdom and experience of experts or people who are closely related to a subject or have knowledge of different aspects of the environment.

To use the Delphi technique, decide on a topic for discussion and invite stakeholders to answer a questionnaire that you have specifically designed for your organization and sustainable innovation project. Or you can use question cards, again designed for your organization, to help groups work together in workshops to capture ideas and intelligence from different people. You can also combine the seven questions future technique above with the Delphi method.

FIGURE 3.4 Sustainable Innovation Roadmap: seven questions canvas

The Delphi technique can be done together, or asynchronously, meaning that people can work at different times and in different places. This makes it a flexible and convenient way to gather insights.

When selecting participants, it's important to be thoughtful about who you invite, as the right mix of people can provide valuable insights that can inform your strategy effectively.

Futures wheel The futures wheel (Figure 3.5) is a tool that helps stakeholders to think about the changes that are happening, or likely to happen, and their impacts. It encourages consideration of the indirect or secondary results of these changes.

To use the tool, start with different central topics such as energy, sustainability, technology, digital AI or any other relevant topic. Invite your team or stakeholders to share their thoughts on the direct results or impacts of the trends that are likely to take place. You can place these answers in a circle outside of the centre.

Next, ask them to think about the indirect results of all the direct results, and place them in an outer circle around the edge.

This approach helps you understand the interconnections and impacts of different trends. You can also use the PESTEL analysis to identify the opportunities and threats, and then think about the indirect results.

Use several different wheels, each with a different topic, as this will help you explore different ideas. The wheels may get quite big with all the different ideas, but this exercise will push your thinking further and make it even more useful.

Scenario planning

Scenario planning is a technique used to mitigate risk, identify opportunities and think through possibilities for your sustainable innovation project. While this approach is often associated with Shell,[64] it can be applied to any industry and help organizations prepare for any potential future scenarios. Here's a step-by-step guide to using your PESTEL insights in scenario planning.

FIGURE 3.5 Sustainable Innovation Roadmap: futures wheel canvas

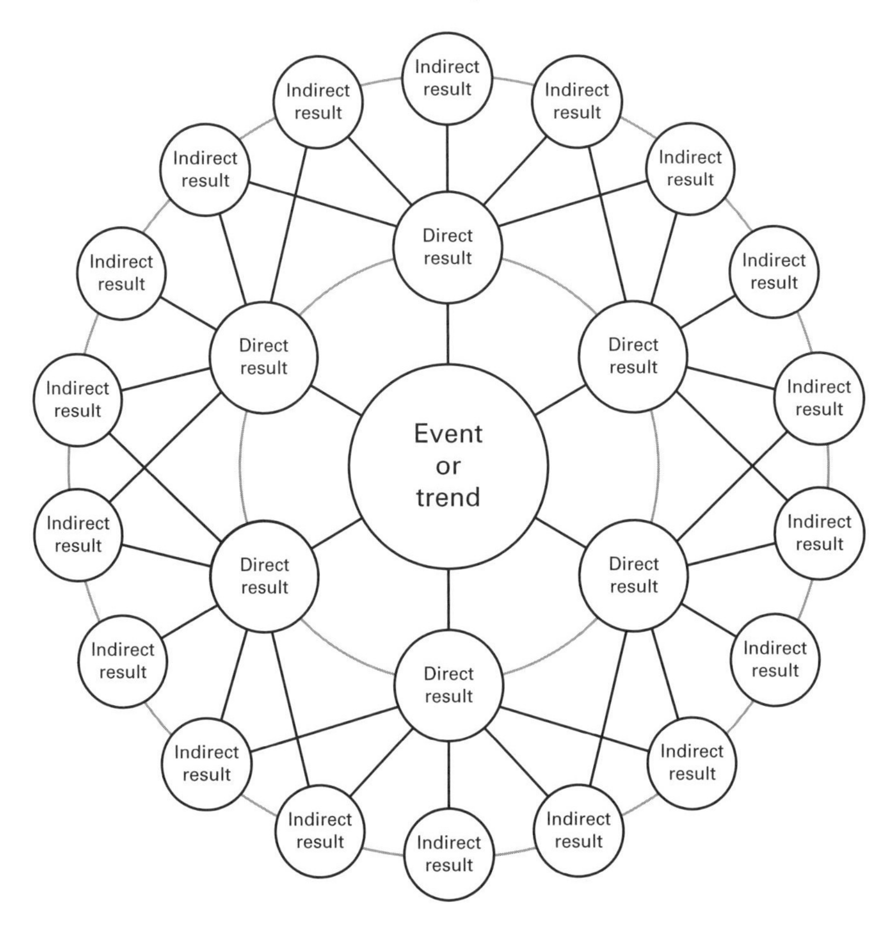

1 **Identify key factors:** Begin by identifying the critical drivers of change for your sustainable innovation, informed by your PESTEL analysis.

2 **Develop scenarios:** Using the scenario planning canvas in Figure 3.6, create diverse scenarios based on different combinations of these key factors. Each scenario should represent a plausible future state, ranging from optimistic to pessimistic outcomes. The futures cone (Figure 3.7) is a useful tool for doing this.

The futures cone is a tool used in scenario planning that helps visualize the range of possible futures ahead. It starts with the present moment at its narrowest point and widens into the future, much like a cone, to represent the broadening spectrum of potential outcomes.

To use the futures cone in scenario planning, begin by identifying the 'preferred future', which sits directly along the cone's central axis. This represents the ideal scenario that a sustainable innovation aims to achieve. The area closest to the axis, within the cone, represents the 'probable futures', i.e. outcomes that are most likely to occur given current trends and data.

Further out from the axis, the cone includes 'plausible futures', which are possible but less likely given what is known today. Beyond that lie 'possible futures', which are theoretically conceivable but not necessarily supported by current evidence or trends. At the edges of the cone are 'preposterous futures', which seem highly unlikely but are not impossible.

3 **Prioritize scenarios:** Once you have several scenarios – i.e. different potential versions of the future – it is important to prioritize them to focus time and resources on those that have the highest levels of importance and certainty, plotting each scenario onto the matrix in Figure 3.8.
4 **Analyse impacts:** Assess how each of your priority scenarios could impact your innovation. Include both opportunities and threats that may arise.
5 **Consider strengths and weaknesses:** Identify internal strengths that could support your innovation and weaknesses that could hinder it.
6 **Develop strategies:** For each scenario, formulate strategies that would enable your organization to respond effectively. This includes contingency plans for adverse scenarios and exploitation plans for favourable ones.
7 **Establish and monitor indicators:** Establish indicators that signal which scenario is becoming more likely over time. Regularly monitor these indicators to adjust your strategies as needed.

FIGURE 3.6 Sustainable innovation roadmap: scenario planning canvas

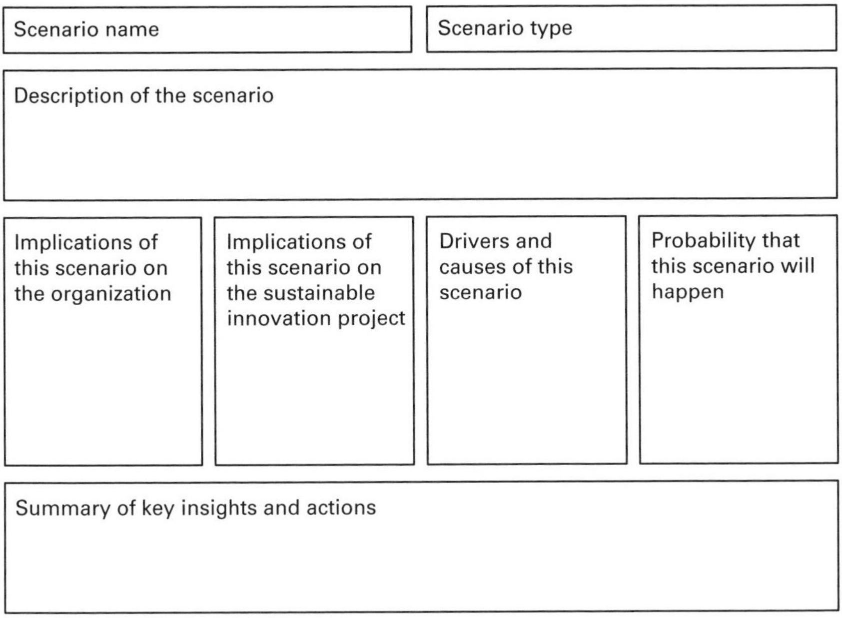

Scenario name
Scenario type
Description of the scenario
Implications of this scenario on the organization
Implications of this scenario on the sustainable innovation project
Drivers and causes of this scenario
Probability that this scenario will happen
Summary of key insights and actions

FIGURE 3.7 Sustainable innovation roadmap: futures cone

Potential
Now
Potential
Everything beyond the present moment
Time
Preposterous
Possible
Plausible
The projected future
Probable
Preferable

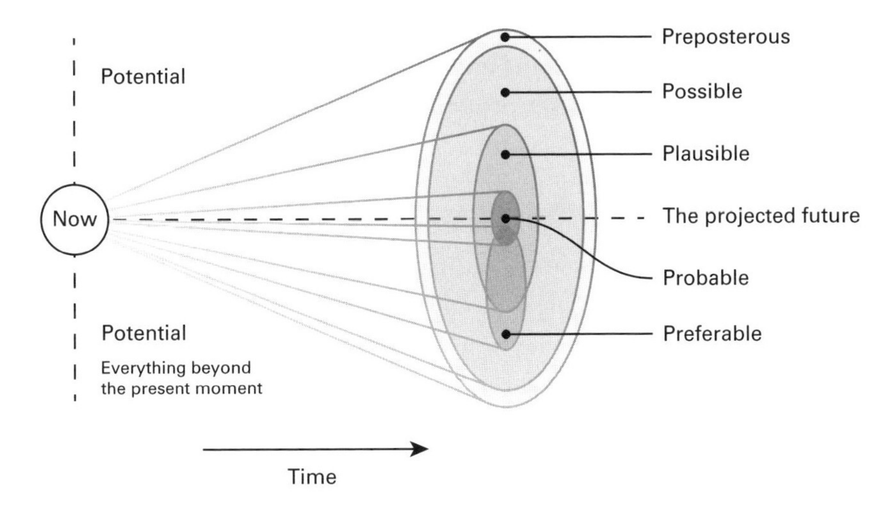

FIGURE 3.8 Sustainable Innovation Roadmap: importance versus certainty matrix

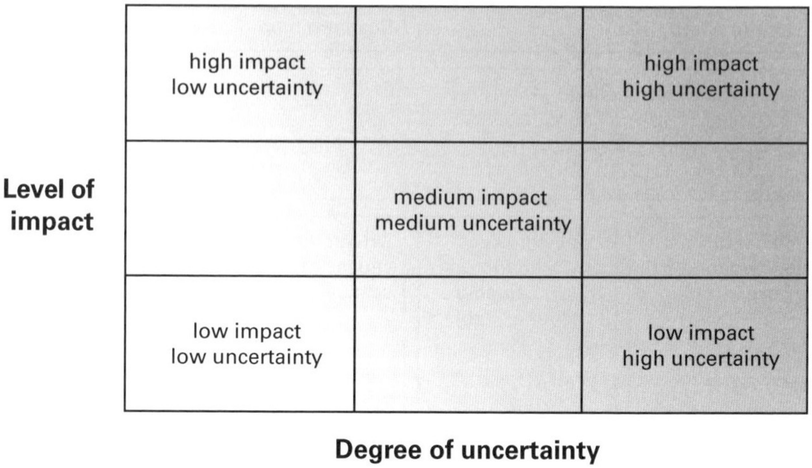

Level of impact			
	high impact low uncertainty		high impact high uncertainty
		medium impact medium uncertainty	
	low impact low uncertainty		low impact high uncertainty
		Degree of uncertainty	

Transformation mapping

Transformation mapping is the final step in the process of horizon scanning. After gathering all the information and insights from PESTEL, stakeholder engagement and scenario planning, it's time to use the priorities from your importance versus certainty matrix and assess how things are likely to play out in the future.

The transformation map (Figure 3.9) is an excellent framework for this purpose. It consists of a future state in the top right-hand corner and all the key aspects for your sustainable innovation project around the edges. Starting from the current state of each aspect, and using your horizon scanning, you can project where your innovation will be at various points in the future. Or you can work in the other direction, and use 'backcasting', which simply means working backwards from your sustainable innovation vision to identify the steps necessary to achieve that future.

Conclusion

The Volatility, Uncertainty, Complexity and Ambiguity (VUCA) of the external environment mean that planning is more necessary than ever to avoid having to react suddenly and falling behind the curve.

FIGURE 3.9 Sustainable Innovation Roadmap: transformation map

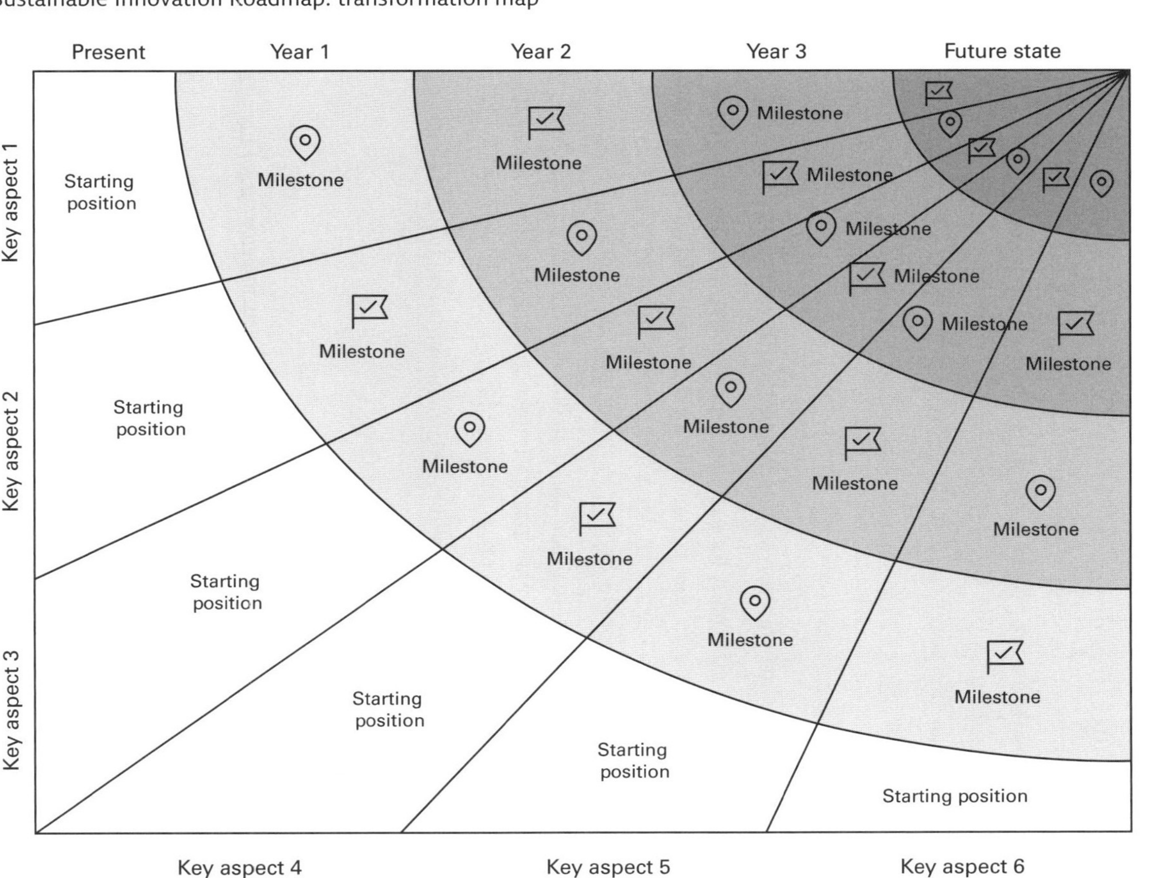

Effective horizon scanning and stakeholder engagement are essential for complex sustainable innovation programmes in technical environments to succeed. They require ongoing focus and updates, attention to weak signals and understanding of the interrelationships and effects that different external factors have on each other. Involving diverse stakeholders helps to ensure that, as far as possible, any limitations or assumptions driven by individual or team biases are offset. By tuning into the signs of change, leaders of sustainable innovation enhance the potential for their programmes to succeed.

ACTION CHECKLIST

- Complete a horizon-scanning activity with your team.
- Create a summary for each key aspect of the scenarios you have prioritized for your sustainable innovation projects. Test and validate these with key stakeholders and experts, using tools such as interviews and the Delphi technique, in conjunction with additional data.
- Build your future vision and map out your current position using the transformation map. 'Backcast' the potential steps for your sustainable innovation projects to get to your future vision from where you are now.
- Update your PESTEL and transformation map regularly, as changes occur.

Notes

1 Bennis, W and Nanus, B (1985) *Leaders: The strategies for taking charge*, Harper and Row, New York

2 Port of Tyne (2024) Our history, www.portoftyne.co.uk/about-us/history/our-history (archived at https://perma.cc/T5VD-62AH)

3 Whitfield, G. Port of Tyne sees turnover fall 22% as it seeks new sources of revenue, *Chronicle Live*, 20 June 2017, www.chroniclelive.co.uk/business/business-news/port-tyne-sees-turnover-fall-13212064 (archived at https://perma.cc/SK8Q-J2NB)

4 Dickinson, K. North East marks the end of an era as the last ever shipment of coal sails from the River Tyne, *Chronicle Live*, 18 February 2021, www.chroniclelive.co.uk/news/north-east-news/coal-shipment-durham-port-tyne-19866501 (archived at https://perma.cc/3M9D-TK25)

5 Griesbach, D (2023) *The Lean Innovation Guide: A proven approach for innovation success*, BIS Publishers

6 Sharpe, B, Hodgson A, Leicester G, Lyon A and Fazey I. Three horizons: A pathways practice for transformation, *E&S*, 2016, 21 (2), 47

7 Intergovernmental Panel on Climate Change (2020) Special Report, www.ipcc.ch/sr15/ (archived at https://perma.cc/6QYX-4SJ3); Serin, E. What technology do we need to cut carbon emissions?, LSE Grantham Research Institute on Climate Change and the Environment, 13 February 2023, www.lse.ac.uk/granthaminstitute/explainers/what-technology-do-we-need-to-cut-carbon-emissions/ (archived at https://perma.cc/8WLW-6JXW)

8 Aguilar, F J (1967) *Scanning the Business Environment*, MacMillan Co., New York

9 Biermann, F, Hickmann, T, Sénit, C A, et al. Scientific evidence on the political impact of the Sustainable Development Goals, *Nature Sustainability*, 2022, 5, 795–800

10 Meng, Q, Su, C-W, Umar, M, Lobonţ, O-R and Manta, A G. Are climate and geopolitics the challenges to sustainable development? Novel evidence from the global supply chain, *Economic Analysis and Policy*, 2023, 77, 748–63

11 European Commission (n.d.) Green infrastructure, environment.ec.europa.eu/topics/nature-and-biodiversity/green-infrastructure_en#:~:text=The%20EU%20Green%20Infrastructure%20Strategy,deliver%20their%20services%20to%20people (archived at https://perma.cc/E59F-FMHC)

12 US Energy Information Administration (2022) Renewable energy explained: Incentives, www.eia.gov/energyexplained/renewable-sources/incentives.php#:~:text=The%20federal%20tax%20incentives%2C%20or,%2D Recovery%20System%20(MACRS) (archived at https://perma.cc/PP2T-DCFZ)

13 International Energy Agency (2023) Trends in charging infrastructure, Global EV Outlook 2023, www.iea.org/reports/global-ev-outlook-2023/trends-in-charging-infrastructure (archived at https://perma.cc/Q4YV-RVLC)

14 Swanson, A and Penn, I. President Biden extends solar tariffs, with major caveats, *The New York Times*, 4 February 2022, www.nytimes.com/2022/02/04/business/economy/solar-tariffs-caveats.html (archived at https://perma.cc/X3AV-LVB6)

15 European Commission (n.d.). Overview of sustainable finance, finance.ec.europa.eu/sustainable-finance/overview-sustainable-finance_en (archived at https://perma.cc/LTW4-9FGV)

16 UN Environment Programme Finance Initiative (n.d.). Sustainable Blue Finance, www.unepfi.org/blue-finance/ (archived at https://perma.cc/8GGA-97UU)

17 Filewood, B. (2023). 'What is conservation finance?', LSE Grantham Research Institute on Climate Change and the Environment, 26th April, www.lse.ac.uk/granthaminstitute/explainers/what-is-conservation-finance/ (archived at https://perma.cc/M66C-UDG4)

18 United Nations (n.d.). Blue Economy Definitions, www.un.org/regularprocess/sites/www.un.org.regularprocess/files/rok_part_2.pdf (archived at https://perma.cc/VX2B-TYXE)

19 United Nations Environment Programme (n.d.). Green Economy, www.unep.org/regions/asia-and-pacific/regional-initiatives/supporting-resource-efficiency/green-economy (archived at https://perma.cc/B3YW-8Y9D)

20 Sidar, C. Suddenly AI: The fastest adopted business technology in history, Forbes, 5 April 2023, www.forbes.com/sites/forbestechcouncil/2023/04/05/suddenly-ai-the-fastest-adopted-business-technology-in-history/ (archived at https://perma.cc/E8ZY-FZ42)

21 Chui, M, Issler, M, Roberts, R and Yee, L (2023) McKinsey Technology Trends Outlook 2023, www.mckinsey.com/capabilities/mckinsey-digital/our-insights/the-top-trends-in-tech#/ (archived at https://perma.cc/AU22-TAYC)

22 McKinsey. What is Web3?, McKinsey & Company, 10 October 2023, www.mckinsey.com/featured-insights/mckinsey-explainers/what-is-web3 (archived at https://perma.cc/BEE2-GA5W)

23 EY Americas (2022) Why it's important to the metaverse an enable of sustainability, www.ey.com/en_us/workforce/the-metaverse-as-an-enabler-of-sustainability (archived at https://perma.cc/XGJ3-4EXE)

24 De Giovanni, P. Sustainability of the Metaverse: A transition to Industry 5.0, *Sustainability*, 2023, 15 (7), 6079

25 World Economic Forum (2023) From Generative AI to sustainable aviation fuel: The top 10 emerging technologies of 2023, www.weforum.org/press/2023/06/from-generative-ai-to-sustainable-aviation-fuel-the-top-10-emerging-technologies-of-2023/#:~:text=These%20include%20flexible%20batteries%2C%20generative,of%20experts%20against%20several%20criteria (archived at https://perma.cc/MQ37-GFHA)

26 Maersk (n.d.) All the way to zero, www.maersk.com/sustainability/all-the-way-to-net-zero#:~:text=Every%20year%20100%2C000%20vessels%20powered,of%20global%20greenhouse%20gas%20emissions (archived at https://perma.cc/RVQ8-VRNG)

27 Maersk (n.d.) Climate change, www.maersk.com/sustainability/our-esg-priorities/climate-change (archived at https://perma.cc/ET2P-6RCG)

28 Ørsted (2022) Ørsted and Maersk sign landmark green fuels agreement, as Ørsted enters the US Power-to-X market, orsted.com/en/media/ news/2022/03/20220310491311 (archived at https://perma.cc/5WEH-8UEE)

29 Hakirevic Prevlkak, N. Maersk secures green fuel supply for 12 methanol-powered boxships, Offshore Energy, 10 March2022, www.offshore-energy.biz/maersk-secures-green-fuel-supply-for-12-methanol-powered-boxships/ (archived at https://perma.cc/PTJ6-CCC8)

30 Vitale, C. Maersk and MAN ES set to retrofit containership, Ship Technology, 22 June 2023, www.ship-technology.com/news/maersk-and-man-es-set-to-retrofit-containership/#:~:text=Danish%20shipping%20company%20AP%20Moller,first%20for%20the%20shipping%20industry (archived at https://perma.cc/M3JD-W56P)

31 MAN Energy Solutions (n.d.) The world's first two-strike methanol engine, www.man-es.com/marine/products/two-stroke-engines/man-b-w-me-lgim (archived at https://perma.cc/G8Y6-7G4R)

32 Maersk (2022) A.P. Moller – Maersk engages in strategic partnerships across the globe to scale green methanol production by 2025, www.maersk.com/news/articles/2022/03/10/maersk-engages-in-strategic-partnerships-to-scale-green-methanol-production (archived at https://perma.cc/L4V4-S5PW)

33 Maersk (n.d.) All the way to zero, www.maersk.com/sustainability/all-the-way-to-net-zero (archived at https://perma.cc/L97S-GRPB)

34 Dogger Bank Wind Farm (n.d.) doggerbank.com/ (archived at https://perma.cc/7G56-GXBR)

35 Offshore Renewable Energy. Quantifying the benefits of tidal stream energy to the wider UK energy system, Catapult Offshore Renewable Energy, 1 June 2022, ore.catapult.org.uk/analysisinsight/quantifying-the-benefits-of-tidal-stream-energy-to-the-wider-uk-energy-system/ (archived at https://perma.cc/XX5A-UY8Q)

36 ELEMENT Project (2023) Summary of the ELEMENT project, https://element-project.eu/news/summary-of-the-element-project/ (archived at https://perma.cc/RV3K-5RXT)

37 Weidenkaff, A, Wagner-Wenz, R and Veziridis, A. A world without electronic waste, *Nature Reviews Materials*, 2021, 6, 462–63

38 IPBES (2019) Summary for policymakers of the global assessment report on biodiversity and ecosystem services

39 Chan, F K S, Griffiths, J A, Higgitt, D, Xu, S, Zhu, F, Tang, Y-T, Xu, Y and Thorne, C R. 'Sponge City' in China—A breakthrough of planning and flood risk management in the urban context, *Land Use Policy*, 2018, 76, 772–78

40 UN Environment Programme (2023) A green wall to promote peace and restore nature in Africa's Sahel, www.unep.org/news-and-stories/story/green-wall-promote-peace-and-restore-nature-africas-sahel-region#:~:text=By%202030%2C%20the%20Great%20Green,beset%20by%20drought%20and%20poverty (archived at https://perma.cc/Z8JS-3AQB)

41 Mondal, B K, Kumari, S, Ghosh, A and Mishra, P K. Transformation and risk assessment of the East Kolkata Wetlands (India) using fuzzy MCDM method and geospatial technology, *Geography and Sustainability*, (2022), 3 (3), 191–203

42 IUCN (2020) IUCN Global Standard for NbS, www.iucn.org/news/europe/202007/iucn-global-standard-nbs (archived at https://perma.cc/3SRC-UTPV)

43 Offshore Renewable Energy Catapult (n.d.) About ORE Catapult, ore.catapult.org.uk/about/ (archived at https://perma.cc/23L9-9SLG)

44 Offshore Renewable Energy Catapult (n.d.) Innovation, ore.catapult.org.uk/what-we-do/innovation/ (archived at https://perma.cc/M5EC-MKL9)

45 Offshore Renewable Energy Catapult (n.d.) ore.catapult.org.uk/ (archived at https://perma.cc/97MJ-5KRT)

46 Lynch, C. The construction industry is getting greener: why, how, and what's changing?, Forbes, 25 August 2021, www.forbes.com/sites/sap/2021/08/25/the-construction-industry-is-getting-greener-why-how-and-whats-changing/?sh=34f4f3e652bc (archived at https://perma.cc/E5UX-N6GZ)

47 Public Law Project. Government 'behind the curve' on AI risks, 13 June 2023, publiclawproject.org.uk/latest/government-behind-the-curve-on-ai-risks/ (archived at https://perma.cc/7HGG-CGME)

48 European Commission (n.d.) Corporate sustainability reporting, finance.ec.europa.eu/capital-markets-union-and-financial-markets/company-reporting-and-auditing/company-reporting/corporate-sustainability-reporting_en (archived at https://perma.cc/Z2CM-ANCF)

49 IFRS (2023) ISSB issues inaugural global sustainability disclosure standards, www.ifrs.org/news-and-events/news/2023/06/issb-issues-ifrs-s1-ifrs-s2/ (archived at https://perma.cc/35UV-PMHB)

50 European Commission (n.d.) European Climate Law, climate.ec.europa.eu/eu-action/european-climate-law_en#:~:text=The%20European%20Climate%20Law%20writes,2030%2C%20compared%20to%201990%20levels (archived at https://perma.cc/5SYM-7MER)

51 European Commission (n.d.) Implementing and delegated acts – SFDR, finance.ec.europa.eu/regulation-and-supervision/financial-services-legislation/implementing-and-delegated-acts/sustainable-finance-disclosures-regulation_en (archived at https://perma.cc/6WSJ-4CUD)

52 PwC (2020) Environmental, social and governance (ESG) in Asia, www.pwc.com/sg/en/asset-management/assets/environmental-social-and-governance-in-asia-awm.pdf (archived at https://perma.cc/Y44F-R3KR)

53 The White House (2022) Inflation Reduction Act Guidebook, www.whitehouse.gov/cleanenergy/inflation-reduction-act-guidebook/ (archived at https://perma.cc/GL44-W6M5)

54 European Commission (n.d.). EU taxonomy for sustainable activities, finance.ec.europa.eu/sustainable-finance/tools-and-standards/eu-taxonomy-sustainable-activities_en (archived at https://perma.cc/EDE3-TKNV)

55 European Commission (2023) EU taxonomy for sustainable activities, finance.ec.europa.eu/system/files/2021-11/211104-ipsf-common-ground-taxonomy_table-call-for-feedback_en.pdf (archived at https://perma.cc/63RS-9NWB)

56 HM Treasury (2021) Fact Sheet: Net Zero-aligned Financial Centre, www.gov.uk/government/publications/fact-sheet-net-zero-aligned-financial-centre/fact-sheet-net-zero-aligned-financial-centre (archived at https://perma.cc/X7G8-ZX4C)

57 Wilson-Otto, G. Regulation is shaping the future of sustainability, Fidelity International, 14 June 2023, www.fidelityinternational.com/editorial/article/regulation-is-shaping-the-future-of-sustainability-836b59-en5/ (archived at https://perma.cc/A34Y-6VQK)

58 Energy, Oil & Gas Magazine. How WUN H2 is uniting heavyweight industry players in the production of green hydrogen, *Energy, Oil & Gas Magazine*, 15 December 2022, energy-oil-gas.com/news/how-wun-h2-is-uniting-heavyweight-industry-players-in-the-production-of-green-hydrogen/ (archived at https://perma.cc/UG5W-8V2B)

59 Siemens (2022) Siemens commissions one of Germany's largest green hydrogen generation plants, press.siemens.com/global/en/pressrelease/siemens-commissions-one-germanys-largest-green-hydrogen-generation-plants (archived at https://perma.cc/MTM7-R3Y3)

60 North, J (n.d.) Horizon scanning, The Big Bang Partnership, bigbangpartnership.co.uk/horizon-scanning/ (archived at https://perma.cc/9NM5-JRGR)

61 US Energy Information Administration (2022) Renewable energy explained: Incentives, www.eia.gov/energyexplained/renewable-sources/incentives.php#:~:text=The%20federal%20tax%20incentives%2C%20or,%2DRecovery%20System%20(MACRS) (archived at https://perma.cc/J7VZ-S26U)

62 Wieckowski, A G. Predicting the Future, *Harvard Business Review*, November–December 2018

63 North, J (n.d.) Stakeholder engagement for innovation success, The Big Bang Partnership, bigbangpartnership.co.uk/stakeholder-engagement-for-innovation-success/ (archived at https://perma.cc/5ZLM-HAP6)

64 Shell (n.d.) Shell scenarios, www.shell.com/news-and-insights/scenarios.html (archived at https://perma.cc/H7JR-DXRR)

04

Green business models and value propositions for technical industries

CHAPTER OVERVIEW

Designing and implementing sustainable business models is paramount to driving long-term success and stakeholder value. This chapter explores the principles and practices of sustainable business models tailored specifically for technical industries. Through real-world case studies, the chapter provides successful examples of business models that align with environmental and societal goals, at the same time as driving profitability and competitiveness.

Contents and Sustainable Innovation Roadmap resources include:

- Crafting value propositions that drive competitive advantage and stakeholder value.
- Designing sustainable business models that align with environmental and societal goals.
- Infrastructure planning and internal processes to align with the sustainable business model.
- Examining successful case studies of sustainable business models in technical industries.

Sustainable innovation value proposition and business models

By integrating your sustainable innovation efforts with your company's core goals and principles, you've laid a strong foundation. The steps covered in Chapters 2 and 3 span from horizon scanning to stakeholder engagement and prioritization of your sustainability focus areas. These steps have been critical in developing a solid understanding of the surrounding factors and key points to consider for your sustainable innovation journey.

The next steps are to consider the value that each of your sustainable innovation programmes and projects will make to your organization and stakeholders, and how you will create that value and operate and maintain them once your innovations become business-as-usual (BAU). This means creating a value proposition and business model for each of your sustainable innovations. The business model, defined as how you organize your programme to achieve your value proposition, will be in two phases. Firstly, the design and build of your asset or product, and then the ongoing operation and maintenance of it. You'll need to ensure that, in addition to being consistent with your organization's purpose and priorities, your innovation achieves the value it is intended to create for users, and your business. This value creation is the central aim of your innovation.

Sustainable innovation as value creation

Chapter 2 defined value as the level of importance, usefulness or worth ascribed to sustainable innovation. It can be quantifiable, such as an investment cost, return on investment figure or growth in market share. In the context of sustainable innovation, value often transcends the tangible and quantifiable to include perceptions and emotions that resonate with stakeholders. Value is not only about the economic benefit but includes the ethical, environmental and social implications that people experience and believe to be significant.

As Hart and Milstein (2003) define it, sustainable value means 'strategies and practices that contribute to a more sustainable world while simultaneously driving shareholder value'.[1] But the perceived or felt value of sustainable innovation can be as impactful as its financial and measurable value. It captures how individuals and communities feel about the innovation's contribution to the greater good, influencing consumer choices, investor decisions and the overall reputation of a company. This kind of value creates a powerful narrative that can drive brand loyalty, attract talent and secure a competitive edge in the market.

Every sustainable innovation needs a sustainable value proposition

Every sustainable innovation needs a sustainable value proposition, that is, a statement of the benefits that the sustainable innovation aims to deliver to the organization and its stakeholders. These benefits need to be well communicated and compelling enough to garner support and buy-in from stakeholders, while also being transparent and realistic about what may or may not be achieved. Value is created through all these benefits working in combination. Of course, what some stakeholders may find valuable may be of no or little interest to others. A single sustainable innovation may require several complementary value propositions, each one addressing the needs and aspirations of a different audience.

Clarity is key

To optimize your sustainable innovation's potential for success, the one non-negotiable is for you to be super clear about the specific nature of the value it is intended to create, and for whom it is being created. Without clarity, all actions, from ideation through to communication, testing, validation, implementation and BAU, will be compromised. Sustainable innovation in technical projects is inherently complex and challenging. It's important to avoid adding to the complexity and challenge by not being completely sure about what you're aiming to achieve.

Fundamental to value creation is the core principle that your sustainable innovation intends to solve a problem that is worth solving, in a way that will be well received. The groundwork you've done from Chapters 2 and 3 will help you to validate where your innovation projects fit within the greater scheme of your organization's ambition and stakeholder priorities. You'll also need to test and validate the specific value proposition for each significant project to check that it does indeed solve the right problem in the right way, and for the right people. As we've seen earlier in Chapter 3, this means going beyond the science, engineering and technical aspects and considering human perspectives and perceptions, too.

The good news is that once the sustainable value proposition is defined well, it makes many elements of the innovation process so much easier. The innovation will be simpler to explain, understand and sell. Investment cases will be clearer. Project teams and suppliers will have more clarity of direction, and be able to execute better and faster. You'll understand what is and isn't working for your sustainable innovation more quickly and transparently, and pivot more quickly if necessary.

Having made the case for a clear, compelling sustainable value proposition, we'll now move on to, first, understanding the concept of value in greater depth and then progressing to some tips and guidance on how to create a sustainable value proposition and business model.

Deconstructing value creation

Achieving sustainable value creation means having clarity on the answers to the following questions.[2]

- **For which stakeholders is value being created?** Consider the number and diversity of stakeholders. Make sure that you prioritize your stakeholders based on factors such as influence over and interest in your sustainable innovation. It's usually impossible to please all the people all the time, so don't try. Identify your most important stakeholders, internally and externally, focus on those and maintain a watching brief on the rest in case anything changes.
- **With whom is value being created?** These are collaborators, mutually beneficial relationships and suppliers.
- **What are the origins of value and the various types of value produced?** Different kinds of value in a business come from various sources. Value can come from both tangible elements

such as physical resources and intangible elements such as brand reputation or company culture. Value comes not just from the traditional financial aspects but also from the social and environmental impact, known as the Triple Bottom Line (referred to in Chapter 1, and also see later in this chapter), along with other kinds of value that might not be as obvious.

- **Where is value being created?** Value can be created in the organization's own activities, through collaborations with others, or within a larger network or ecosystem, society and the environment.
- **When is value being created?** Unlike traditional activities that often focus on immediate profits, sustainable innovation projects require thinking about the long term. This means organizations need to balance their short-term gains with long-term success and sustainability. The effects of sustainable innovation in technical environments often show over a long period, which is why it's important to think about the entire lifespan of a product or service. Leaders of sustainable innovation should focus on creating assets and making products that last longer and can be easily repaired, updated or reused. They should also design with a full lifecycle in mind, from creation to disposal, to make the most of materials and minimize waste.
- **What conflicting interests are there, and how might these be managed?** Leaders of sustainable innovation often encounter situations where different goals related to sustainability might conflict with each other. It's important to understand what these conflicts are and find ways to handle them. When creating sustainable value, a business works with various people and groups who may have different aims and interests, leading to tension. These conflicts can happen over time, as what creates value now might not do so in the future, or across different places, where what works in one location may not in another. Leaders of sustainable innovation should actively look for these potential conflicts and either solve them or find ways to manage them to continue creating value sustainably.

CASE STUDY

Value creation at Maersk

Maersk's Sustainability Report 2023 explains that it's a purpose-driven company which aims to deliver 'better, simpler, and more reliable outcomes for our customers – improving life for all by integrating the world'.[3]

The report goes on to explain who Maersk creates value for:

> **Our customers**
>
> We aspire to provide truly integrated logistics for 100,000+ customers' supply chains.
>
> **Our people**
>
> We keep our people safe and engaged while offering equitable and interesting career paths.
>
> **Society**
>
> By integrating global logistics, we improve the flow of goods and materials that sustain people, businesses, and economies the world over and contribute to improved quality of life and prosperity.
>
> **The planet**
>
> Our industry is a significant contributor to global greenhouse gas emissions, and we are committed to realising net zero supply chains by 2040.
>
> **Shareholders**
>
> In our transformation to become the global integrator of container logistics, we continue to innovate and grow shareholder value.

The report also contains more detail on how the company creates value for each of these different stakeholders.

JOHN GOOD GROUP: AN EXAMPLE OF THE TRIPLE BOTTOM LINE

A key element of a sustainable business model is a value proposition that provides measurable environmental and/or social value with economic value.[4]

The Triple Bottom Line (TBL) is a concept that broadens a business's focus on the financial bottom line to include social and environmental considerations. It's sometimes described as being like a three-legged stool where one leg represents financial performance, another stands for social responsibility and the third for environmental stewardship. These are often referred to as the 'three Ps': People, Planet and Profit (or Performance). Together, they support a sustainable business strategy that aims for long-term value creation:

Environmental responsibility focuses on enhancing the efficiency with which resources are utilized, ensuring that the resources are used responsibly. The goal is to avoid any negative impacts on the environment, including harmful emissions, striving instead to promote environmental well-being.

Social accountability involves ensuring the health and safety of all involved, adhering strictly to legal frameworks and respecting the rights of employees, stakeholders and individuals. It is also about upholding ethical principles and avoiding social harms, all while contributing to the enhancement of societal well-being.

Economic viability concentrates on improving cost-effectiveness and expanding profits and business opportunities. It underlines the importance of bolstering resilience and stability in operations, along with reducing risks. Collectively, these practices aim to improve economic well-being for the company and its stakeholders.

Some organizations have adopted the TBL to align with today's consumer values, attract and retain employees and mitigate risks associated with environmental and social factors. By measuring performance across these three areas, businesses can aim for profitability while also contributing positively to the world around them.

One such example is John Good Group, a sixth-generation family business in the UK, with companies involved in maritime, supply chain logistics and storage, travel management and property. John Good Group describes its commitment to the TBL as:

> **People**: As part of our philosophy, we partner with like-minded teams who are passionate about maximising the positive impact their work has on their sector and society as a whole.
>
> **Planet**: Our work reflects a deep-rooted respect for the planet on every level, from small neighbourhoods and communities to cities, countries and the global environment.
>
> **Performance**: Monitoring and evaluation are at the core of how we operate, which means that every project is assessed on its performance and continually improved to bring top results.[5]

Stakeholder perceptions of value

Stakeholder value is multifaceted, and leaders of sustainable innovation need to consider a range of factors to fully understand and cater to their stakeholders' needs and preferences. Some ways in which stakeholders derive value from a sustainable innovation are:[6,7]

- **Financial**: how stakeholders perceive the cost savings or financial benefits they receive for their investment, considering both immediate and long-term expenses.
- **Functional**: the practical utility and performance quality of the innovation, and how well it does what it's supposed to do.
- **Emotional**: how stakeholders feel about or experience the innovation.
- **Social**: how the innovation can enhance a stakeholder's social image, brand or reputation, or how it enables them to connect with others.
- **Learning**: the innovation's ability to stimulate learning and discovery, providing new experiences or knowledge.
- **Situational**: how certain conditions or situations can make the innovation more valuable or useful to the stakeholder.

Of course, stakeholders may benefit from an innovation that provides two or more of these factors at the same time.

Effective stakeholder engagement, as discussed in Chapter 3, is a must to ensure that your organization truly understands stakeholders' needs and aspirations. It is also essential that your organization clearly communicates the value that your sustainable innovation projects deliver for your stakeholders, using language and approaches that resonate with them. Communicating value is almost as important as delivering value in many cases, because stakeholders are unlikely to benefit if they are unaware of what your activities are and why you are doing them.

Value proposition 'fit'

When shaping your value proposition statement for your sustainable innovation, you're aiming for value proposition 'fit'.[8] This means that your stakeholders can readily understand how the proposed project will effectively address their key goals or challenges in a way that they find acceptable. The more direct the fit, the easier it will be for your stakeholders to support your idea.

Your value propositions should always be developed, tested and validated with your key stakeholders. Research by Laukkanen and Tura (2022) shows that perceptions of stakeholder value created by the organization can be significantly different from how stakeholders view the value proposition.[9] Always check for alignment.

THE NEED FOR VALUE PROPOSITION 'FIT': A FINANCIAL EXAMPLE

Research by Giannetti et al (2022) recommends that leaders of sustainable innovation and investors need to reconsider how they define 'value' to align with the UNSDGs.[10] The traditional ways by which companies are evaluated for their worth – like profit margins or share prices – don't accurately represent how sustainable a company's practices are. They need to be replaced with new strategies that are based on clear, science-backed data and systemic thinking. By using a mining company in Brazil as a case study, Giannetti et al highlight the importance of concrete, scientific methods to evaluate a company's sustainability. They argue for the use of quantifiable indicators that can guide investors towards companies that not only offer financial returns but also contribute to the UNSDGs. The goal is to encourage investments in genuinely sustainable businesses.

CASE STUDY

Port of Tyne: pivoting with purpose

In Chapter 3, we discovered how a rapidly changing energy landscape forced the Port of Tyne to confront a watershed moment that tested its resilience and adaptability. This port, once heavily reliant on coal, had to reinvent its value proposition and business model to thrive in a new era of sustainability.

Mark Stoner, Chief Financial Officer, joined the Port of Tyne in 2018 to work with the management team on leading the business to recovery and growth. He says:

> In the mid-2010s, the Port of Tyne found itself at a crossroads. With a business model heavily dependent on coal – a resource that had previously

> driven 70 per cent of its income – the port was on the brink of obsolescence as the coal industry collapsed. This drastic shift in market dynamics needed an urgent and strategic response. The port faced a challenging financial situation, burdened with debt and a need for a revitalized business plan. The leadership was tasked with a formidable challenge: to reinvent the port's identity and leverage its assets for future markets.

Mark explained that the port's management embarked on a rigorous evaluation of emerging markets and the opportunities within the decarbonization and net zero landscape. Their strategy focused on identifying competitive advantages, particularly geographical ones, which could be utilized to enter and compete in new markets successfully.

An immediate opportunity arose with the transition from coal to biomass. Power generation customers shifted to importing wood pellets, a change the port was well equipped to accommodate using its existing infrastructure. This pivot showcased the port's agility and willingness to adapt to changing environmental demands.

The port's management also made tough choices to cease non-profitable ventures, notably its UK distribution business. This decision was a difficult but important step to stem financial losses and concentrate on core competencies.

Post-restructuring, the focus was on three primary areas where the port excelled:

1. Landlord business
2. Product handling
3. Passenger services and safe navigation

The Port of Tyne concentrated on these strengths, targeting niche markets where it could command a competitive edge.

Mark said:

> The Port of Tyne opted to capitalize on specific niches, such as its core handling capabilities and geographical deep-water positioning. Instead of competing with larger ports on every front, it sought to compete in specialized areas. We crafted a new value proposition, grounded in strong ethical and environmental values. The port aligned itself with customers who shared these values, ensuring mutually beneficial relationships. This alignment also meant selecting customers willing to acknowledge and recognize the value the port brought to the table.

Securing the initial customers was critical in establishing credibility. By demonstrating capability and reliability, the port started to attract more customers, creating a virtuous cycle of trust and business growth. The new business brought in by these customers enabled the Port of Tyne to reinvest in equipment and infrastructure, preparing for sustainable growth. This growth was not just in financial terms but also in building a reputation as a forward-looking, environmentally responsible entity.

Today, the Port of Tyne stands as a testament to adaptability and resilience. By acknowledging the need for change, being open to global shifts, leveraging competitive advantages, prioritizing opportunities and creating a compelling story, the port not only survived a potential downfall but also positioned itself as a leader in sustainable business practices.

Mark summarizes his step-by-step process as follows:

Step 1: assess your current position

Acknowledge where you stand, without excuses. Recognize that the status quo must change.

Step 2: scan for challenges and opportunities

Look ahead to identify global trends and challenges that may impact your business in the next 10 to 20 years.

Step 3: identify your strengths

Determine what competitive advantages your business has, such as location, skills or relationships, that can be leveraged for long-term success.

Step 4: craft a compelling story

Create a clear and convincing value proposition. This is crucial to gain customers, secure financing and get team buy-in for the change.

Step 5: prioritize and act

Focus on the most promising opportunities, and don't overlook easy wins that can show quick progress and convince stakeholders of the chosen direction.

The case of the Port of Tyne is a compelling example of transformation and strategic redirection. It demonstrates the importance of being agile in transforming the value proposition and business model in response to global trends.[11]

MAERSK'S NEW METHANOL-POWERED VESSELS CREATE NEW, SUSTAINABLE VALUE PROPOSITIONS

In 2023, Maersk ordered six mid-sized container vessels – all having dual-fuel engines able to operate on green methanol. Yangzijiang Shipbuilding Group were commissioned to build the six 9,000 TEU vessels for delivery in 2026 and 2027.[12] Based on the information published about these vessels,[13,14,15,16] we could describe the value proposition as:

For customers and partners

Maersk's innovative fleet of green methanol-powered vessels leads the maritime industry into a sustainable future. Its 9,000 to 17,000 TEU capacity vessels, crafted by world-class shipbuilders, are a testament to its commitment to environmental stewardship and technical excellence. Delivered between 2024 and 2027, these ships are not only a step towards Maersk's green transformation but also a promise to its customers of cleaner, more efficient and responsible shipping solutions. With this ground-breaking initiative, Maersk ensures that customers' cargo not only reaches global markets but also supports a sustainable supply chain that reduces greenhouse gas emissions significantly, paving the way for a greener tomorrow.

For investors and stakeholders

Maersk stands at the vanguard of the shipping industry's green transition with its order of methanol-enabled vessels capable of operating on green fuel. These vessels represent a strategic investment in its fleet's future, aligning with global sustainability goals and responding to the increasing demand for environmentally friendly shipping options. Maersk's partnership with Equinor ensures a steady supply of green methanol, demonstrating its commitment to long-term sustainability and innovation. As Maersk continues to pioneer the use of green fuels, it invite others to join it in setting a new industry standard for sustainability and technological advancement, ensuring a competitive edge and a future-proof business model.

STEPS FOR DRAFTING A SUSTAINABLE VALUE PROPOSITION STATEMENT

Following these steps will help you to draft a sustainable value proposition statement for your innovation:

Step 1: understand your stakeholders

Before you begin, identify who the value proposition is for. Understand their needs, challenges and the benefits they seek. In a technical environment, this could include stakeholders such as customers, investors, communities and internal team members.

Step 2: define the innovation

Clearly articulate what the innovation is and the technology behind it. Ensure that you can explain it in simple terms, avoiding jargon that could confuse non-technical stakeholders.

Step 3: highlight sustainability

Detail how your innovation contributes to sustainability. This could be through energy efficiency, reduced waste or better use of resources. Provide tangible examples or projections.

Step 4: explain the benefits

Articulate the direct benefits of your innovation. How does it improve on current solutions? Focus on long-term advantages, such as cost savings, performance improvements and environmental impact.

Step 5: differentiate

Clarify what sets your innovation apart from others. In a complex technical environment, differentiation could be green solutions created by advanced technology, unique application of existing technology, use of new materials or integration with other systems.

Step 6: connect to bigger goals

Link your value proposition to broader goals and trends, such as global sustainability targets or industry standards. Show how your innovation aligns with these objectives.

Step 7: refine your message

Craft a concise statement that brings all the above points together. It should be clear, compelling and easy to understand, including for stakeholders without technical expertise.

Step 8: test and iterate

Share your value proposition with a small, diverse group of priority stakeholders. Gather feedback and refine your statement as needed to ensure it resonates and communicates the intended value.

Step 9: implement

Once finalized, consistently use your value proposition in all communication about the innovation project to ensure a unified message.

An excellent value proposition is not just a description. It's a statement of the unique contribution your innovation will make to the market, society and the planet. It should be a blend of practicality and inspiration, grounded in technical feasibility and lifted by a vision of a more sustainable future.

What if your organization's drive for sustainability is ahead of what your key stakeholders value?

In many organizations, stakeholders lobby organizations for sustainable innovation and investment. However, in some cases, the organization's drive for sustainability is more ambitious than that of some of their important stakeholders, such as their communities, shareholders, employees or management team. Leaders of sustainable innovation can influence their stakeholders by understanding why people resist change and by developing strategies to help overcome those barriers.

The reasons why people resist change are often emotional rather than rational. We make decisions based on emotion, and how we want to feel, and then justify those decisions with logical argument.[17] Fear of the unknown and uncertainty about the future can make people uncomfortable with change. Change can also feel like a loss of control over one's environment or job, leading to resistance. Changes that alter how people perceive their role or status can be threatening, and people often prefer familiar routines and may resist altering their habits. Sometimes, resistance can simply be down to bad timing; for example, if it comes at a time of high stress or workload, it can be unwelcome. Unconscious bias and assumptions can also influence judgement, too, as we've seen in Chapter 2. Plus, if the advantages of sustainable innovation aren't clear, some stakeholders may not support it. Resistance is also likely if there's a mistrust in the leadership or the motivations behind the change.

When you're asking stakeholders to support your sustainable innovation, you're asking them to place their trust in you and in your plans. Trust is created when stakeholders feel that the person or team that they are dealing with are credible, reliable, authentic and motivated by good intentions.[18] Some values-led strategies that help to generate trust and influence support for your sustainable innovation where there is resistance to change are:

- Understanding the change curve, which maps the emotional journey people go through when dealing with change, and why people resist change.
- Engaging and collaborating with stakeholders, involving them in the innovation process. We'll cover this in more detail in Chapters 10 and 11.
- Emphasizing the aspects of your innovation that align most closely with your stakeholders' interests, goals and objectives.
- Identifying early adopters and influential advocates to generate early support for your plans and lead by example. By targeting and convincing these key individuals or groups to support your sustainable innovation, you create early success stories that demonstrate the benefits and feasibility of your initiatives. Influential advocates, such as industry leaders or public figures, can amplify your message. Their endorsement can lend credibility and visibility to your innovation. They can also act as role models, showcasing the adoption of your innovation in a practical setting. This strategy can create a bandwagon effect, encouraging more people to join in simply because they see people they respect doing so.
- Where appropriate, offering incentives for adopting sustainable options.
- Sharing the challenges and potential achievements of your sustainable innovation with transparency and integrity to help build trust. Never unintentionally or otherwise engage in greenwashing, a deceptive practice of misleadingly portraying innovations as eco-friendly, but base your sustainability (and all other) claims on evidence. Be transparent about risk and show clearly how you plan to mitigate it.

Value creation, delivery and capture through sustainable business model innovation

A sustainable business model is the plan for how the organization will deliver the sustainable innovation value proposition to stakeholders and generate the intended financial and non-financial returns on investment.

> Sustainable business models aim to create significant positive impacts or significantly reduce negative impacts for the environment and/or society through changes in how the organisation and its value network create, deliver, and capture value,[19] or change their value propositions. (Bocken et al., 2014)[20,21]

The business model's purpose is to create, deliver and capture value from the sustainable innovation. If the value proposition is even slightly off-target, then the business model will be too.

While sustainable technical innovation projects, new assets, products and services are not generally perceived to be 'businesses' in the traditional sense, they do require a business model. This is because a business model outlines the strategy for how an organization will create, deliver and capture value. Even for innovations that prioritize sustainability, there must be a clear plan for how they will reach customers, achieve market viability and ensure financial sustainability.

It is helpful to consider the business model framework as consisting of tangible objects, actors, resources, processes and results designed to generate, distribute and acquire economic, social and environmental benefits throughout the entire lifespan of the sustainable innovation. The Business Model Canvas is a good tool to use, adding social environmental costs and benefits into the original canvas, and the planet as a key stakeholder.[22] The Triple Layered Business Model Canvas is also an excellent option.[23] You could also use the Sustainable Innovation Roadmap Business Model Framework (BMF) in Figure 4.1.

The Sustainable Innovation Roadmap BMF is useful for planning, reviewing and improving sustainable innovation in technical

FIGURE 4.1 Sustainable Innovation Roadmap Business Model Framework (BMF)[24]

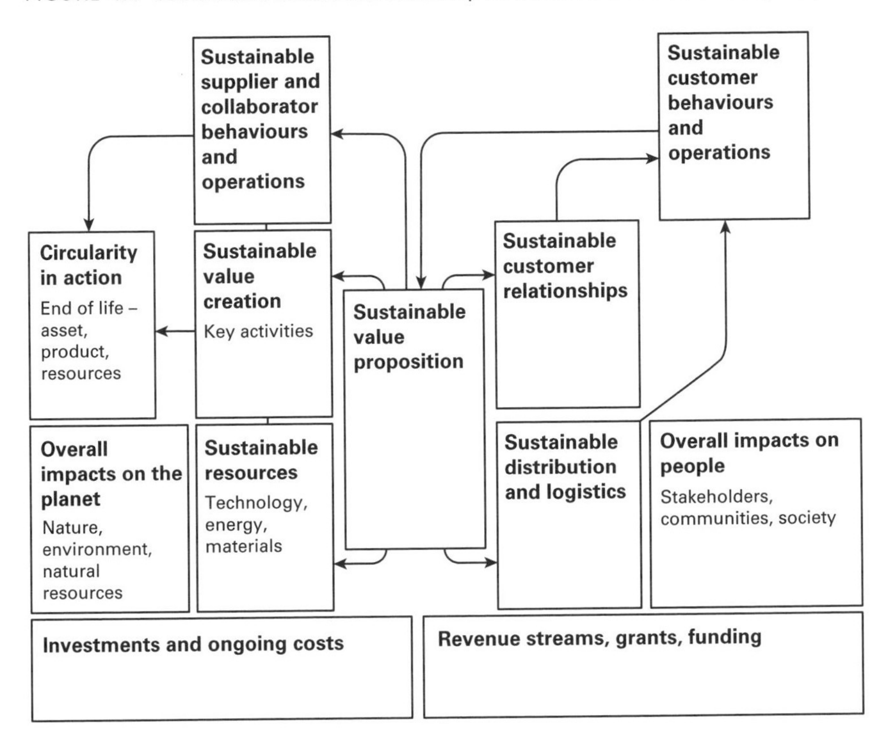

environments. Each innovation ideally will have its own business model. This will need to be nested into the organization's overall business model where it is neither a standalone nor an intentionally disruptive project designed to break out of the organization's current modus operandi.

An effective business model for sustainable innovation projects is essential to attract investment, manage resources effectively and scale the innovation. It helps innovation leaders to articulate the value proposition more comprehensively, identify key partners and activities, define customer segments and stakeholder audiences, and establish revenue streams. Without the business model framework, even the most promising technical innovations can struggle to transition from concept to reality, failing to make the intended impact.

For instance, renewable energy projects, while technologically advanced and environmentally beneficial, still need to determine how they will generate income, whether through selling clean energy, carbon credits or related services. Similarly, any new sustainable asset or product must find its market fit and understand its cost structure and distribution channels to be successful.

Which comes first, the value proposition or the business model? It depends. If the business is a new start-up or an organization experiencing strategic drift, the value proposition is the starting point and the business model configured around it. If there is an existing successful sustainable business model, the innovation needs strategic fit within it, preferably leveraging existing resources, processes and relationships. That said, a dynamic business model is essential. It needs to constantly adapt to shifts in demand, new technology and emerging competitive landscapes. This requires an ongoing innovation process in the continuous search for new ways to sustainably capture customer value that are challenging for others to replicate.

It's important to map out the business model for your sustainable innovation. Doing so has several advantages:

- **Narrative and financial consistency:** A well-constructed business model combines the sustainable innovation project's story with its finances, helping everyone understand how your project makes money despite uncertainties and complexities. It's a communication tool that helps you to spot areas for improvement.
- **Comparison:** You can compare different potential business models that aim for the same goal but do it differently to compare your options.
- **Versatility:** As explained earlier, business models aren't just for the whole organization; they can be for individual assets, products or different projects, reflecting unique needs and market situations.

HOW TO USE THE SUSTAINABLE INNOVATION ROADMAP BMF

While you can use the Sustainable Innovation Roadmap BMF (Figure 4.1) in any order that works best for your organization and project, a suggested approach is:

1. Start in the centre of Figure 4.1 by adding your sustainable value proposition. You probably have a few slightly different value propositions to meet the needs of different stakeholder groups. Include the most important ones here.

2. Capture information about the customers that your sustainable value proposition will support, and how you intend to build and maintain sustainable relationships with those customers in the *Sustainable customer relationships* box.
3. Add how your innovation project will help your customers to operate more sustainably in the *Sustainable customer behaviours and operations* box.
4. Now consider the key activities that you will need to carry out to deliver your value proposition. Break each key activity down into individual, granular steps. Identify how you will achieve each of these activities sustainably. Summarize your plans in the *Sustainable value creation* box. Add circularity aspects to the *Circularity in action* box.
5. Next identify the resources that you'll use to achieve your value proposition and carry out the key activities in the previous step. Include materials, energy, transport, packaging, people, technology and so on. Work through your sustainability approach for each of these resources. Note your plans in the *Sustainable resources* box. At the same time, consider how you will integrate circularity into the resources used in your sustainable business model. Include your approach in the *Circularity in action* box.
6. Consider the support, actions, behaviours, relationships and operations that you'll need from your suppliers and collaborators to achieve your sustainable value proposition in line with your activity, resource and circularity plans. Capture your conclusions in the *Sustainable supplier and collaborator behaviours and operations* box.
7. Define the overall impacts that your business model decisions in steps 1–6 above will have on the planet (i.e. nature, environment, natural resources) and on people (i.e. stakeholders, your local communities, society more widely). Add your summary to the boxes for *Overall impacts on the planet* and *Overall impacts on people* respectively.
8. Identify the investments and ongoing costs you will need to make your value proposition and business model work, and the income sources (e.g. revenue from sales, grants, funding, investment, donations) they are likely to attract. Add your notes to the *Investments and ongoing costs* and *Revenue streams, grants, funding* boxes at the bottom of the framework.

9 Consider your sustainable business model in the round, and make any adjustments needed. Iterate these to explore several sustainable business model options until you have one that best balances your requirements for sustainability, affordability, time and risk. We will explore the tension and balance between these requirements in Chapter 7.

NEW BUSINESS MODELS NEEDED FOR INFRASTRUCTURE INNOVATION FOR RESILIENT AND SUSTAINABLE URBAN AREAS

Goal 11 of the UNSDGs aims to make cities inclusive and sustainable. A study by Walsh et al (2022) asserts that current infrastructure business models don't fully value their long-term benefits, and new models are needed that focus on local innovation, address specific issues and improve how we evaluate infrastructure investments.[25]

The study highlights that traditional business models for infrastructure projects do not adequately capture the full scope of economic, social and environmental benefits that such projects can provide. It suggests a need for innovative business models that:

- Focus on urban or regional areas at a local scale, where innovation can have the most significant impact.
- Address specific challenges or needs, making it clear what innovations are necessary.
- Rethink the economic cost-benefit analysis traditionally used to assess infrastructure investments, which may be flawed and not reflect the true value.

Business model innovation in this context means creating frameworks that encourage collaboration among multiple stakeholders to develop and implement infrastructure solutions that deliver wide-ranging values to all involved parties. This approach involves mapping out the diverse benefits of infrastructure over time and space to understand the interconnections and the broader impact. By acknowledging a wider array of value propositions, innovative business models aim to attract more stakeholders and identify alternative funding and financing mechanisms to support sustainable infrastructure projects. These new models are critical for unlocking investment in situations where conventional methods have been unsuccessful, supporting the achievement of resilience, sustainability and regeneration goals.

CASE STUDY

Sustainable business model: agrivoltaics

Agrivoltaics merge agriculture with photovoltaic solar power to utilize land more efficiently. This sustainable innovation business model promotes dual land use for simultaneous agricultural production and electricity generation. The examples here explore the development and implications of this new, sustainable business model, analysing its components such as value proposition, activities, resources, user benefits and financials.

Examples and applications

- In China, agrivoltaics are extensively used in the form of solar greenhouses, which covered up to 1.96 million hectares or 30 per cent of the country's horticultural area as of 2021. These greenhouses harness solar energy while providing a conducive environment for growing popular vegetables like cucumbers, melons, tomatoes and peppers, even in colder months.
- Germany's agrivoltaics project in Donaueschingen-Aasen uses vertically mounted panels over 14 hectares. This design employs bifacial solar modules that allow for grazing space below, demonstrating the dual land-use concept effectively.
- Japan's approach to agrivoltaics includes a vertically designed system in Nihonmatsu City. Installed by the Institute for Sustainable Energy Policies and Ryoeng Co., Ltd, it offers pasture space between the solar panel rows, available for local livestock farming.
- France is testing an agrivoltaics system for growing hops, which can reach up to 8 meters in height. The French government actively supports agrivoltaics through competitive contracts, recognizing the potential of this technology to contribute to the country's renewable energy goals.

Reasons for use

The reasons for the adoption of agrivoltaics vary. In regions like China, where agricultural land is at a premium, solar greenhouses allow for year-round crop production despite harsh winters. In Europe, where environmental sustainability is a significant goal, countries like Germany and France see agrivoltaics as a way to meet renewable energy targets without sacrificing agricultural productivity.

Value proposition

The core value of agrivoltaics lies in its dual-use nature, which not only increases the productivity of land but also contributes to sustainability goals. By installing solar panels over agricultural land, the model provides a renewable energy source while still allowing for crop cultivation or grazing below. This setup creates a microclimate that can improve plant growth and reduce water needs due to less evaporation.

Operational activities

Operational activities include the installation of solar panels designed to coexist with agricultural operations. This requires sophisticated tracking systems to optimize sunlight exposure for both energy generation and crop growth. It also includes managing the interaction between the panels and the plants, ensuring that shading patterns are beneficial rather than detrimental to the crops.

Resources

The resources critical for this business model are bifacial solar panels, tracking systems, transparent polymers for panel construction and AI software for managing light exposure and panel movement. It also involves specialized agricultural knowledge to understand how different crops interact with the altered microclimate.

User benefits

Farmers benefit from a new revenue stream from energy generation, alongside their traditional agricultural income. The protection provided by solar panels can lead to higher yields and potentially extend the growing season. In some cases, the solar infrastructure can offer shelter to livestock, increasing animal welfare and productivity.

Financials

While the initial capital expenditure for agrivoltaic systems is higher than traditional solar installations, the dual revenue stream has the potential to offset these costs. The higher upfront costs stem from the need for specialized equipment that can accommodate agricultural activities. However, the long-term operational costs can be lower due to the synergistic benefits of combined energy and agricultural production. Maintenance can be streamlined as the presence of vegetation beneath the solar panels can result in less soil erosion and, therefore, less sediment accumulation on the panels, reducing cleaning costs.

The financial success of agrivoltaics is dependent on several factors, including local climate conditions, crop types and the technical specifics of the system installed. The model's profitability is contingent on finding the right balance between energy generation and agricultural productivity.

The model can also face challenges due to the higher costs associated with complex system designs and the need for modifications in agricultural machinery and practices.

Despite these challenges, agrivoltaics stands as a promising development in the sustainable business arena, offering a practical solution to the food–energy nexus. As this model continues to evolve, it could play a significant role in transforming both the renewable energy and the agricultural sectors, leading to more resilient and sustainable practices.[26]

Adapting the business model framework for sustainable innovation in technical environments

There are some important, specific considerations to build into the business model design process for sustainable innovations in technical environments. These are:

- **Building circularity into the business model from the earliest stage possible.** As discussed in Chapter 3, the further ahead we try to forecast, the less detailed and accurate our thinking is likely to be. But it's important to design in circularity principles from the very beginning to reduce potential remediation requirements, to create a no-harm, or positive, legacy at the end of the innovation's useful life. A different business model is likely to be needed for each phase of the innovation's lifecycle once it is live.

EVOLUTION OF SELLAFIELD'S SUSTAINABLE BUSINESS MODEL

The site known today as Sellafield began its life as Windscale, part of the UK's post-war atomic bomb project. The site underwent significant changes over the years, with each phase requiring a different business model to adapt to the changing priorities and uses of the facilities.

Initially, the Windscale site was integral to the UK's atomic weapons programme, with the Windscale Piles constructed to produce plutonium. This phase of the site's life required a business model focused on research and development, as well as on the production of materials for national defence.

Later, the site expanded and transitioned into the Calder Hall nuclear power station, the world's first to export electricity on a commercial scale to a public grid. This evolution broadened Windscale's business model to include energy generation and commercial activities, marking a shift from purely military production to civilian energy services.

In 1971, the site was renamed Sellafield, and its activities expanded further into nuclear fuel reprocessing, which continued until 2022. The business model during this period centred on the processing of nuclear material and the handling of nuclear waste, services that were critical as the nuclear energy industry grew.

Currently, Sellafield is in its decommissioning phase, which represents another significant shift in its business model. Decommissioning involves complex processes to dismantle and safely dispose of or store nuclear materials and infrastructure. This phase is characterized by a focus on safety, environmental protection and long-term waste management, with business activities including clean-up operations, waste processing and the development of new technologies to manage the decommissioning process effectively.

The decommissioning of the Windscale Pile One chimney, a notable feature of the Sellafield site, is an example of the complex and unique challenges involved in this phase. The process requires innovative approaches and technologies to dismantle structures safely and cost-effectively, given the radiological considerations and the historical significance of the site.

Throughout these transitions, the ownership and management of the site have also evolved, reflecting the changing nature of the nuclear industry in the UK. The site is now managed by Sellafield Ltd on behalf of the UK Nuclear Decommissioning Authority, indicating a business model that is closely aligned with government oversight and public accountability.[27,28,29,30,31]

- **Consider what will happen when your innovation becomes a legacy asset.** We will explore sustainably innovating existing legacy assets in detail in the next chapter. For now, it's important to understand that today's exciting, state-of-the-art technical solution will be tomorrow's legacy asset.

- **Plan in continuous improvement to keep your innovation relevant, functional and fit for purpose.** Build in the capability to future-proof it as far as is reasonably practicable, by designing in the space for add-ons and extensions as capacity and other requirements grow.
- **Factors such as complexity of value and levels of societal and environmental responsibility are of even greater importance for sustainable innovations in technical environments, especially where there is some kind of public sector involvement.** The business model needs to be created and evolved with these needs in mind. Plus, these needs are likely to change over time. The business model will need to interact with other stakeholders' business models. It is important to build in the flexibility to be able to do this from the outset, based on what can reasonably be done with the information available at the time.

FROM LEGACY ASSET TO INNOVATIVE 'SUPER SEWER'

This large-scale infrastructure project is an example of how former innovations, such as London's sewers, evolve into legacy assets that require modernization to meet current needs. It also shows how the new asset has been designed to accommodate future needs, and incorporate societal and environmental responsibility, partly fuelled by public sector involvement.

The Thames Tideway Tunnel in the UK, also known as London's 'Super Sewer', is a significant upgrade to the capital's outdated sewer system, originally designed during the Victorian era for a much smaller population.

The original London sewer system, constructed in the late 19th century by Sir Joseph Bazalgette, was an engineering marvel of its time, designed to prevent cholera outbreaks by diverting waste away from drinking water sources. However, the system was not built to withstand the current population's demand, leading to frequent overflows into the River Thames.

To address this issue, the Thames Tideway Tunnel was conceived as a modern solution to intercept and manage the overflow of sewage and stormwater, prevent pollution in the Thames and accommodate the needs of London's growing population. Spanning 25 kilometres, the tunnel follows the course of the river from west to east and captures combined sewer overflows from several sites.

Financially, the Tideway project represents a significant investment, utilizing a unique financing model where a consortium, rather than the utility company Thames Water, carries the costs. This arrangement aims to mitigate financial risks and minimize the impact on taxpayer bills, with oversight from the regulatory body, Ofwat, to ensure value for money. The UK government has also committed to providing financial support if costs exceed a certain threshold.

In terms of construction, the Tideway Tunnel incorporates numerous innovations to enhance efficiency and productivity. These include the utilization of tunnel-boring machines for excavation, an innovative 'vortex' design to manage the flow of effluent and a dedicated Innovation Forum to seed fund and develop new technologies. The Tideway Tunnel has incorporated future-proofing by designing in space for add-ons and extensions as capacity requirements grow.

The project has also been mindful of environmental and health and safety considerations. For instance, it uses the river for transporting materials to reduce road traffic and improve air quality, and it has established an employers' project induction centre to increase safety awareness among workers.

The Tideway Tunnel is set to capture at least 94 per cent of sewage overflows into the Thames each year, significantly improving the river's ecological state and the health of recreational water users. It demonstrates the evolution of London's infrastructure, from the innovative Victorian sewers to a modern 'Super Sewer' designed to meet the challenges of the 21st century.[32]

Assessing the sustainability impact of a business model

Once you have developed your framework for your sustainable business model, it's helpful to carry out a Sustainability Impact Assessment to evaluate and measure its environmental, social and economic effects. A Sustainability Impact Assessment has sub impact assessments, which are:

1 **Environmental:** This aspect assesses the potential environmental effects of the activity. It considers factors such as resource consumption, pollution, habitat disruption and greenhouse gas

emissions. The aim is to identify and mitigate any negative impacts on the environment.

2. **Social**: The Social Impact Assessment evaluates how the activity may affect the well-being of communities, individuals and society. It looks at issues like social equity, cultural heritage, human rights and quality of life. The goal is to enhance positive social outcomes and address any adverse effects.
3. **Economic**: This part examines the economic consequences of the activity. It assesses aspects such as job creation, economic growth, cost-benefit analysis and financial sustainability. The aim is to ensure that the activity contributes to economic prosperity without causing harm.
4. **Health**: The Health Impact Assessment focuses on the potential health effects of the activity. It considers factors like air and water quality, access to healthcare and lifestyle changes. The goal is to safeguard public health and well-being.
5. **Cumulative**: This considers the combined effects of multiple activities or projects on sustainability factors. It helps identify synergies and trade-offs between different initiatives.
6. **Biodiversity**: For activities that may affect ecosystems and wildlife, this assessment looks at preserving biodiversity and habitat integrity.
7. **Strategic**: This is a broader assessment used for policies, plans and programmes rather than individual projects. It ensures that sustainability considerations are integrated into higher-level decision-making processes.

There will always be trade-offs. An action that may improve or help one component could compromise another. You'll need to create balance, while working to clear objectives and priorities.

A process for achieving balanced trade-offs

The sustainable innovation process is iterative. While your initial materiality analysis in Chapter 2 provided a broad direction from a wider organizational perspective, it's time to refine this analysis to align closely with the unique aspects of your sustainable value proposition and business model. You'll need to customize the prioritization to fit the particulars of your project, sharpening your materiality analysis to ensure that it's appropriately tailored and relevant. The following process is useful to help achieve balance across different sustainability and other components, and multiple stakeholders, each with their own priorities and agendas:

1. **Restate your objectives:** Define the objectives of your organization and the innovation. What are you trying to achieve, and what are the key outcomes you want to prioritize? Reminding yourself of your well-defined objectives will help to guide your assessment process.
2. **Take a systems thinking approach:** Focus on the interactions and relationships between the components that make up the entirety of your innovation and its impact, and map this out visually. Systems thinking acknowledges that all parts of a system are interconnected, and changes in one part of the system can have intended and unintended consequences on other parts of the system. Identify these interconnections and feedback loops that can either stabilize the system (negative feedback) or lead to exponential growth or decline (positive feedback). Highlight leverage points, i.e. where small changes could lead to significant improvements in the overall innovation. A helpful resource on how to do this for infrastructure projects is Bouch's (2020) 'Systems Approach to Infrastructure BMs'.[33]
3. **Involve priority stakeholders:** Involve priority stakeholders from various groups, including environmental experts, community representatives, economists and relevant experts. Their perspectives are invaluable for understanding the different aspects of sustainability and ensuring a balanced solution.
4. **Collect data:** Gather available data on each component of your sustainable innovations, including environmental, social, economic

and health impacts. Use both quantitative and qualitative data sources to get a rounded and comprehensive understanding.

5 **Do your analysis:** Use a structured approach that combines different criteria and their weights to rank and prioritize alternatives, such as the one in Figure 4.2. Select the most balanced solution by considering all relevant factors:
   - Specify relevant criteria for evaluating and comparing your options.
   - Determine weights for the criteria, representing their relative importance to your organization and stakeholders.
   - Rate the options based on the criteria, resulting in a ranking to inform your decision-making.

We'll go into greater detail on analysis and decision-making in Chapters 8 and 9.

6 **Iterate:** Review and refine the assessment as new information becomes available or as project details evolve.

FIGURE 4.2 Decision analysis grid

		Option 1		**Option 2**		**Option 3**		**Option 4**	
Criterion	Weight	Rating	Value	Rating	Value	Rating	Value	Rating	Value
Quality	1	5	5	3	3	1	1	1	1
Ease of implementation	4	1	4	4	16	1	4	1	4
Support	2	2	4	1	2	1	2	3	6
Total			13		21		7		11

Rating Scale:
1 = Doesn't meet criterion, 2 = Somewhat meets criterion, 3 = Meets criterion, 4 = Slightly exceeds criterion, 5 = Significantly exceeds criterion

Conclusion

Every sustainable innovation needs a clear, compelling set of complementary value propositions that resonate with key stakeholders. These sustainable value propositions provide the core of your internal and external communications about the innovation, investment cases and more, as well as being at the centre of the business model you design to implement and maintain it. Your sustainable value propositions need to reflect the multiple types of value that matter the most to your stakeholders, and in language that connects with them.

Sustainable business model innovation requires the identification of all the key components that will achieve the value proposition, and how they will combine to realize its intended benefits. It's important to carry out a Sustainability Impact Assessment on the proposed business model, and work through any trade-offs that you may need to make using an informed, intentional process.

ACTION CHECKLIST

- Define your key stakeholders and your innovation value proposition statements for each.
- Test and validate your value proposition statements with your key stakeholders to ensure they align with stakeholders' needs and expectations.
- Design the optimal business model to realize the value propositions, carrying out a Sustainability Impact Assessment and working through any trade-offs that you may need to make.
- Keep you value proposition statements and business model updated as changes occur and as new information and insights emerge.

Notes

1 Hart, S L and Milstein, M B. Creating sustainable value, *AMP*, 2003, 17, 56–67

2 Manninen, K, Laukkanen, M and Huiskonen, J. Framework for sustainable value creation: a synthesis of fragmented sustainable business model literature, *Management Research Review*, 2023, 47 (1)

3 Maersk (n.d.) Sustainability at A.P. Moller – Maersk, www.maersk.com/sustainability/reports-and-resources (archived at https://perma.cc/HS2Y-PDMK)

4 Laukkanen, M and Tura, N. The potential of sharing economy business models for sustainable value creation, *Journal of Cleaner Production*, 2020, 253, 120004

5 John Good Group (n.d.) https://johngoodgroup.co.uk/ (archived at https://perma.cc/CN5B-MBV2)

6 Sweeney, J C and Soutar, G N. Consumer perceived value: The development of a multiple item scale, Journal of Retailing, 2001, 77 (2), 203–20

7 Laukkanen, M and Tura, N. Sustainable value propositions and customer perceived value: Clothing library case, *Journal of Cleaner Production*, 2022, 378, 134321

8 Osterwalder, A, Pigneur, Y, Bernarda, G, Smith, A and Papadakos, T. (2014) *Value Proposition Design: How to create products and services customers want*, 1st edn., Wiley

9 Laukkanen, M and Tura, N. Sustainable value propositions and customer perceived value: Clothing library case, *Journal of Cleaner Production*, 2022, 378, 134321

10 Giannetti, B F, Scarpelin, J, Di Agustini, C A, Paranhos, M A H L, Lozano, P A, Agostinho, F and Almeida, C M V B. Perceived value versus real value: Why can investors in sustainable companies fail in their mission?, *Cleaner Production Letters*, 2022, 3, 100020

11 Interview with the author, 16 February 2024

12 Maersk (2023) Maersk orders six methanol powered vessels, www.maersk.com/news/articles/2023/06/26/maersk-orders-six-methanol-powered-vessels (archived at https://perma.cc/FJ7Z-JQ8B)

13 Maersk (2023) Maersk signs landmark green methanol offtake agreement, significantly de-risking its low-emission operations in this decade, www.maersk.com/news/articles/2023/11/22/maersk-signs-landmark-green-methanol-offtake-agreement#:~:text=A.P.%20Moller%20,25 (archived at https://perma.cc/E7QD-9VKP)

14 Maersk (2022) A.P. Moller – Maersk continues green transformation with six additional large container vessels, www.maersk.com/news/articles/2022/10/05/maersk-continues-green-transformation#:~:text=Copenhagen%2C%20Denmark%20%E2%80%93%20A,TEU (archived at https://perma.cc/U7S2-7C97)

15 Maersk (2023) Maersk to pioneer first container vessel conversion to methanol dual-fuel engine, www.maersk.com/news/articles/2023/06/21/maersk-to-pioneer-first-container-vessel-conversion-to-methanol-dual-fuel-engine#:~:text=In%202021%2C%20we%20ordered%20the,fuel%20methanol%20engines (archived at https://perma.cc/MG6R-NA8G)

16 Maersk (2023) Equinor and Maersk partner up to ensure continued green methanol supply for the world's first methanol-enabled container vessel, www.maersk.com/news/articles/2023/09/08/equinor-and-maersk-partner-to-supply-first-methanol-enabled-container-vessel#:~:text=Copenhagen%2C%20Denmark%20%E2%80%93%20A,the%20first%20half%20of%202024 (archived at https://perma.cc/UM7N-QEME)

17 Mlodinow, L (2022) *Emotional: How feelings shape our thinking*, Penguin Random House

18 Maister, D H, Galford, R and Green, C (2021) *The Trusted Advisor*, 20th Anniversary Edition, Free Press

19 Kaplan, S (ed.) (2012) Business Models 101: Creating, delivering, and capturing value, in *The Business Model Innovation Factory: How to stay relevant when the world is changing*, 1st edn, Wiley

20 Bocken, N M P, Short, S W, Rana, P and Evans, S. A literature and practice review to develop sustainable business model archetypes, *Journal of Cleaner Production*, 2014, 65, 42–56

21 Laukkanen, M and Tura, N. Sustainable value propositions and customer perceived value: Clothing library case, *Journal of Cleaner Production*, 2022, 378, 134321

22 Osterwalder, A and Pigneur, Y (2010) *Business Model Generation: A handbook for visionaries, game changers, and challengers*, Wiley

23 Joyce, A, Paquin, R and Pigneur, Y. The triple layered business model canvas: a tool to design more sustainable business models, ARTEM Organizational Creativity International Conference, 26–27 March 2015, Nancy, France

24 North, J (2024) Figure created by the author

25 Walsh, C L, Glendinning, S, Dawson, R J, O'Brien, P, Heidrich O, Rogers, C D F, Bryson, J R and Purnell P. A systems framework for infrastructure business models for resilient and sustainable urban areas, *Frontiers in Sustainable Cities*, 2022, 4, 825801

26 Department of Energy (2022), Market Research Study: Agrivoltaics, https://science.osti.gov/-/media/sbir/pdf/market-research/seto---agrivoltaics-august-2022-public.pdf (archived at https://perma.cc/9CTL-TZMP)

27 Law, E. What is Sellafield?, NDA Blog, 7 September 2018, https://nda.blog.gov.uk/what-is-sellafield/ (archived at https://perma.cc/P5D9-ZUBQ)

28 GOV.UK (n.d.) Sellafield Ltd, www.gov.uk/government/organisations/sellafield-ltd/about (archived at https://perma.cc/K459-ELLD)

29 Subramanian, S. Dismantling Sellafield: The epic task of shutting down a nuclear site, *The Guardian*, 15 December 2022, www.theguardian.com/environment/2022/dec/15/dismantling-sellafield-epic-task-shutting-down-decomissioned-nuclear-site (archived at https://perma.cc/FL3E-6RB2)

30 Temperton, J. Inside Sellafield: how the UK's most dangerous nuclear site is cleaning up its act, WIRED, 17 September 2016, www.wired.co.uk/article/inside-sellafield-nuclear-waste-decommissioning (archived at https://perma.cc/RS5V-F8DZ)

31 Chemeurope.com (n.d.) Sellafield, Encyclopedia, www.chemeurope.com/en/encyclopedia/Sellafield.html (archived at https://perma.cc/ZJ2H-FU3U)

32 Tideway (n.d.) The Tunnel, https://tideway.london/the-tunnel (archived at https://perma.cc/2PSV-UP8W)

33 Bouch, C (n.d.) Systems thinking to develop alternative infrastructure business models, University of Birmingham, www.lboro.ac.uk/media/wwwlboroacuk/content/systems-net/downloads/160615%20-%20SysNET%20presentation%20v3.pdf (archived at https://perma.cc/3E6L-LFG5)

05

Working with legacy assets and creating sustainable, future-ready solutions

CHAPTER OVERVIEW

Chapter 5 tackles the topic of sustainable asset management. It focuses on updating existing infrastructure and designing new assets with sustainability at their core, as these will set the stage for the future. It presents the application of circular economy principles to drastically cut waste and optimize resource use, adopting approaches that facilitate the reuse of materials. The chapter also addresses the challenge posed by outdated assets that can impede broader innovation, outlining strategies for managing legacy technology and the crucial maintenance of data. It emphasizes the necessity for a clear vision, value proposition and business model in retrofitting these assets, and considers the implications for technology, buildings and infrastructure. The significance of systems thinking is highlighted, coupled with actionable tips. Additionally, the chapter provides practical checklists to assess and enhance the energy efficiency of buildings, and to move towards circular methodologies.

The challenge of legacy assets

Legacy assets are outdated but still operational technology, buildings, machinery or infrastructure that an organization continues to use, even though more up-to-date and often better alternatives are available. These legacy assets are often deeply integrated into the organization's operations, making replacement or upgrading complex and potentially disruptive. Buildings, machinery and infrastructure may have historical significance or may simply be aged and in need of renovation or repurposing to meet current needs and standards. Legacy assets can also include natural resources or landscapes that have been inherited from past generations, often carrying the burden of previous exploitation or contamination that requires current management and remediation efforts. For instance, brownfield sites, once industrial areas, now repurposed for commercial or residential use, can require extensive clean-up and redevelopment. This involves advanced engineering techniques to ensure they are safe and sustainable for future use. Technological innovation plays a key role in assessing, detoxifying and revitalizing these spaces, turning legacy burdens into community assets.

An additional challenge with legacy assets can be their long lifespans, which means they can continue emitting CO_2 for decades, during and after their useful lifespan. A typical cement plant may operate for 40 years, an aircraft for 27 years and a building for 80 years. These long-lived assets can significantly contribute to global temperature increases, water stress and biodiversity loss if not managed or updated to reduce their carbon footprint.[1]

Legacy assets can be a significant barrier to the transition to more sustainable practices and other developments that have a dependency on those legacy assets. It's important to list, review and prioritize your organization's legacy technology, buildings, infrastructure and other assets. Map out any interdependencies between them and any potentially limiting factors they may create for your other sustainable innovation plans.

CAPACITY CHALLENGES OF THE UK'S NATIONAL GRID

The UK's National Grid faces significant challenges that are preventing the integration of renewable energy sources due to its current capacity limitations. In 2022, the *Financial Times* reported that some renewable energy developers can experience up to a decade's delay in connecting new capacity to the grid, jeopardizing the UK's net zero target for 2050.[2]

The government's ambitious goals to expand renewable energy significantly depend on private developments, which are hindered by the grid's insufficient capacity and legacy assets. This is especially problematic in Wales, where farmers and landowners interested in producing renewable energy are constrained by the limited grid connections.[3]

Across the UK, the grid's infrastructure, designed for centralized power stations, struggles to accommodate the decentralized nature of renewable sources like solar farms and rooftop panels. The challenge lies in upgrading the grid to handle distributed generation and the financial implications of such an upgrade. The National Grid, a privatized company, faces regulatory constraints from UK regulator Ofgem on the charges it can pass to consumers, limiting investment potential.

The design of the electricity grid hasn't kept up to date with the advancements in renewable energy. As the UK transitions away from traditional power stations, the grid must adapt to smaller, decentralized energy sources like solar farms and rooftop panels, which are essential for leveraging the full benefits of renewable energy.

To achieve a zero-carbon grid, the UK needs to upgrade the grid's capacity to distribute excess electricity nationwide. This is not a straightforward task. It requires a joint effort from network operators, governmental bodies and owners of transmission lines to successfully upgrade the grid's infrastructure.

The assets we create today are the legacy assets of the future. As new requirements, and new solutions emerge, it's essential that we innovate the legacy assets we have now to create and deliver value more sustainably. Weighing the economic and sustainability costs and returns of building anew against the benefits of maintaining or enhancing existing assets, while also anticipating the resilience and future requirements of these legacy assets, can be complex and challenging. It's important to strategize the handling of existing investments and financial commitments, particularly when emerging technologies, materials and regulations could invalidate or devalue assets by decreasing demand or suitability.

New and increasing risks are also a concern for buildings, infrastructure and technology assets. For example, cybercrime globally is increasing and becoming more and more sophisticated. The systems used to manage these assets need frequent reviews and upgrades to increase protection and adapt to new threats to ensure resilience of operations, maintenance of data integrity and preservation of reputation. Ironically, in some cases, the fact that organizations have not updated and networked all their systems provides some protection in the event of a cyberattack, allowing operations to continue. For example, research by Akpan et al (2022) shows that 'The maritime sector, which until now was considered safe due to the lack of Internet connectivity and the isolated nature of ships in the sea, is showing a 900 per cent increase in cybersecurity breaches on operational technology as it enters the digital era.'[4]

Legacy technology can create significant issues. Outdated hardware and software may lead to cybersecurity risks, but it can also affect critical systems, harming operational resilience and service delivery. Legacy technology can be a barrier to data access, interoperability, GDPR compliance, data streamlining and digital transformation, getting in the way of the development of more innovative IT services.[5] Lack of data consistency across legacy systems within and between organizations can prevent or hinder effective, real-time, or timely data sharing and analysis. Maintenance of legacy systems can be expensive, and over time fewer and fewer people may have the skills and experience to manage and fix them, as well as creating manual workarounds to connect them to new services. Outdated systems can cost more in terms of time, cash and energy to run.

But moving away from legacy technology can be costly. The investment case will need to include factors such as dual running costs when transitioning to new systems, data cleansing and transfer, system deletion time and costs, training and development for new processes and systems once the new technology is in place. At some stage, however, there is likely to be a tipping point when the cost and risk of maintaining the current system outweighs the investment in a new solution. It is generally better to pre-empt this, plan and budget for it in advance.

Gartner identifies seven potential options for innovating legacy IT systems, ranked by the simplest to implement first.[6]

1 Encapsulate

Put the core parts of your software, its data and functions, safe and separate from other functions. Then improve or extend them to add new features and make them available via an API (application programming interface). An API is a set of rules and tools that lets different software programs share data and functionalities in a standardized way.

2. Rehost

   Move your application exactly as it is from one computer or server setup to another, for instance from a physical server to a virtual or cloud one.

3. Replatform

   Replatforming is the same as rehosting, only with small changes made to the software code so that it can work effectively in its new environment.

4. Refactor

   Clean up and organize the code to make it neater and more efficient but without changing the functionality of the program. This helps to fix any underlying issues and inefficiencies.

5. Rearchitect

   Fundamentally change how the code is structured, and how it operates, so that you can take advantage of newer technologies and opportunities.

6. Rebuild

   Start from scratch to make a new version that does the same things as the old one but with newer technology and improved features.

7. Replace

   Get rid of your old software application completely and replace it with a completely new one that better meets your organization's current needs and requirements.

The appropriate strategy to choose will depend on the level and importance of the transformation you need, combined with the costs, time, resource investment and potential risks of changing versus staying the same.

Beware of data deterioration

Protecting or replacing technology assets can be important from the perspective of data preservation and conservation. Data loses its quality, relevance and integrity over time. This data deterioration, also referred to as 'data decay', takes place for several reasons. The media used for the storage itself may degrade, leading to data loss, potentially creating short-term issues as well as denying future generations access to heritage and learning.

MICROSOFT INNOVATES DATA STORAGE TECHNOLOGY

Microsoft's 'Project Silica' is an innovation aimed at addressing the issue of data loss, especially for long-term data storage, using advanced quartz-glass-based storage technology.

Information is stored inside a piece of quartz glass by using extremely fast lasers to make very tiny, precise marks at different levels and angles within the glass. These marks are so small that they're measured on the scale of nanometers – one billionth of a meter – and the process involves creating a three-dimensional pattern that can be read to retrieve the stored information.

This technology is excellent for keeping records safe and in their original condition for a long time, for potentially up to 1,000 years.

As we create more and more digital information, our usual ways of storing it can't keep up; they can't hold enough, and they don't last long enough. Project Silica is potentially a new way to store huge quantities of data that will last a long time without needing to be replaced or upgraded.[7,8]

Legacy assets and constant change

The horizon-scanning approach and tools in Chapter 3 are just as relevant for leading the sustainable innovation of legacy assets as they are for planning brand-new projects. They can help your organization and its stakeholders to recognize and prepare for future influences on your legacy assets to help you maintain their functionality and relevance and extend their useful life sustainably.

Change is constant, which means that information that was once correct can become old or wrong as situations change. Plus, as new technology comes out, the old way of keeping information might not work with the new systems, or it might not be the right kind of information needed now. When it comes to sustainably innovating legacy technology assets, it's essential have a strategy for preserving and updating their associated data. We will explore how data and technology facilitate sustainable innovation in Chapter 6.

CASE STUDY

Decommissioning Sellafield: changing legacies through time

As we saw in Chapter 4, Sellafield, a nuclear site in the UK, has a history stretching back to the 1940s and 1950s, originally established for purposes quite different from its current function. It has evolved from producing nuclear materials for weaponry to generating electricity and then to reprocessing nuclear materials from around the globe. This evolution has left Sellafield as a complex legacy site; its decommissioning wasn't a consideration from the start.

Today, there's a growing focus on decommissioning in the nuclear industry. The challenges Sellafield faces are unique but not isolated; they mirror those at other global nuclear sites undergoing decommissioning. These challenges – and the opportunities they present – are not just for Sellafield but for the wider nuclear decommissioning industry and its supply chain.

Frank Allison is the founder and CEO of FIS360 Ltd, the organization that developed and delivered Sellafield's Game Changers open innovation programme. He explains:

> By 2016 it became apparent that Sellafield alone couldn't address all the challenges of decommissioning and that a significant shift in mindset would be needed. We worked extensively with them to help them understand the benefits of open innovation and to demonstrate how opening up their challenges to a cross sector network could dramatically improve their ability to find the technology that was so crucially needed. Once we'd established a willingness to collaborate externally on challenges, we worked in close partnership with Sellafield to devise a bespoke open innovation programme called Game Changers.[9]

The overriding priority for Sellafield, and Game Changers, is always safety. Achieving excellence in decommissioning efforts is also a key objective.

Sellafield aims for transformation – making operations safer, reducing environmental and financial costs, and involving the workforce in this change.[10]

Game Changers brings diverse thinking from various sectors,[11] fostering innovation by learning from the experiences of others, such as those in water, oil, gas or chemicals, who have faced similar issues.

The programme also demonstrates Sellafield's thorough search for global solutions and expertise to tackle its unique challenges.[12]

We'll go into more detail about the Sellafield Game Changers programme in Chapter 10, which focuses on creating a sustainable innovation ecosystem.

Systems thinking for the sustainable innovation of legacy assets

Systems thinking,[13] an approach for understanding and addressing complex challenges that we explored in the previous chapter, is particularly relevant when innovating legacy technology and infrastructure. It involves seeing the world more holistically and recognizing that all parts are interconnected, leading to consequences because of these connections, not despite them. In practice, this means looking at legacy systems not just as isolated entities but as integral components of a larger organizational ecosystem.

Any legacy asset, whether it's an outdated technology system, an established infrastructure or high-value, complex piece of equipment, doesn't operate in isolation. It's part of a larger system that encompasses current technologies, employee skills, organizational processes, customer interactions, stakeholder impact and more. Understanding how this asset fits into and interacts with the wider system is crucial, especially when planning sustainable innovation upgrades, retrofits or rebuilds.

Legacy assets typically have deep integration with other systems and processes. For instance, an old software system might be tied in with various databases, analytics tools and customer service interfaces. Another example is the upgrade of a city's water system. Existing pipes, pumps and reservoirs are often closely integrated with the city's urban layout, other utilities such as the electricity grid and public services such as roads and transportation. These interconnections mean that the asset plays a significant role in daily operations, influencing everything from data flow, day-to-day running and employee workflows. Careful planning is needed to ensure that, while new, more sustainable and efficient solutions are implemented, the existing service continues without major interruptions.

When legacy assets are innovated, the impacts can ripple through the entire system. Operational disruptions are a common initial outcome, as the introduction of new technology or processes can temporarily slow down or complicate established ways of working. Employees, customers and stakeholders, used to the legacy system, may face a learning curve with the new setup, needing a period of adjustment.

Compatibility is another critical issue. The sustainable innovation must work seamlessly with existing systems. If not, it could lead to a chain reaction of required changes, escalating the project's scope and complexity. This situation can also bring about financial considerations, as the cost of implementing a new system extends beyond the mere purchase price. It includes training expenses, potential downtime and the integration of the new system into the existing infrastructure.

Innovating a legacy asset is not without risks. These can range from data loss and security vulnerabilities during the transition to the possibility that the new system may not fully meet the operational needs or expectations. Such risks necessitate careful planning and robust risk management strategies.

Finally, introducing new technologies or processes often leads to cultural shifts within an organization. Especially in cases where the legacy asset is deeply embedded in the organization's routine, replacing it requires not just a physical change but also a shift in mindset and culture.

Therefore, when contemplating changes to a legacy asset, leaders of sustainable innovation are advised to take a comprehensive view, considering how these changes will affect the entire system. A successful transition from a legacy asset to a more sustainable and innovative solution requires a nuanced understanding of these interdependencies and a strategic approach that encompasses technical, financial, human and cultural aspects.

When applied to sustainable innovation, systems thinking can help to build a future where organizational actions are resilient and considerate of social and environmental impacts. To apply systems thinking in practice, especially for sustainable innovation, leaders will benefit from developing a team culture of critical thinking, digging deeper into issues, seeing the connections between different elements, expanding time and spatial horizons of thinking and considering multiple perspectives. This approach helps to understand the far-reaching implications of changes made to legacy systems, ensuring that any innovations are sustainable, efficient and aligned with the broader goals of the organization.

A common challenge in sustainable business model innovation is ensuring that changes in business activities positively impact the wider system's sustainability. It's important to use systems thinking principles early in the business model development process, to avoid uncovering significant issues later. This includes creating a clearly defined scope, facilitating collaboration and integrating key principles such as interconnections, causal relations, feedback loops and system changes. Manage and avoid overcomplication and overwhelm by focusing on material and specific elements, such as key stakeholders.[14]

SYSTEMS THINKING IN ACTION: AVIATION INNOVATION

Systems thinking in sustainable aviation innovation considers aircraft, airports and air traffic management as interdependent. It includes the concept of 'conscious aircraft' – planes equipped with self-monitoring and predictive maintenance capabilities, contributing to sustainability by potentially reducing the carbon footprint through optimized fuel usage and less frequent need for part replacements, in addition to the growing interest in sustainable aviation fuels.

Airports have become digital hubs, enhancing efficiency in managing passenger flows, baggage and cargo, which aligns with sustainable practices by reducing waste and energy consumption. Real-time data facilitates more effective operations, conserving resources. Digital connectivity also extends to improving the passenger experience, offering personalized services that can lead to more efficient travel and less environmental impact.

However, integrating new technologies into the existing infrastructure brings challenges, including the coexistence of varying generations of technology. Addressing these challenges head-on with an ecosystem-wide approach is crucial for achieving a seamless transition towards more sustainable aviation practices. This digital evolution in aviation aims to revitalize the industry and ensure it thrives on sustainable principles that reduce climate impact and build resilience and adaptability.

Aviation's future also could include the integration of autonomous systems such as drones and personal air mobility vehicles, which present new sustainability considerations in terms of airspace management and environmental impact.[15]

Sustainable value propositions and business models for legacy assets

Chapter 4 showed the importance of developing sustainable value propositions and business models for innovation projects. In the context of legacy infrastructure, buildings, technology and assets, this involves creating strategies that recognize their unique characteristics and potential for adaptation.

Legacy assets often come with restrictions on how they can be modified, which require creative solutions for integrating sustainable practices. Examples of these restrictions could be planning or other regulations, preservation orders and technical considerations. We'll explore how to identify and deal with constraints in Chapter 7. In addition, there may be more stakeholders interested in a legacy asset, especially if it has cultural or historical significance, requiring a more inclusive approach. New assets can be designed from scratch with sustainability in mind, while legacy assets need retrofitting and adaptation, which can be both a challenge and an opportunity for innovation. Integrating modern technology into legacy assets for sustainability often requires custom solutions to avoid altering the asset's character. Financing models for legacy assets may leverage grants and incentives for preservation and sustainability that are not available for new assets.

Owing to their inherent constraints and opportunities, creating sustainable value propositions and business models for legacy assets requires a different approach compared to new assets (see Figure 5.1).

Process for creating sustainable value propositions and business models for legacy assets

1 *Assessment and analysis*

- **Legacy asset audit**

  Conduct a thorough assessment of legacy assets to understand their condition, historical significance and potential for adaptation or reuse. Complete a detailed assessment of the asset to identify its strengths, weaknesses, opportunities and threats (SWOT analysis). Identify quick wins as part of your opportunity analysis (also using the quick win checklist below).

- **Regulatory review**

  Understand the legal and regulatory framework governing your legacy assets, especially for heritage buildings, to determine what changes are permissible.

- **Market research**

  Analyse market demands and trends to identify potential opportunities for repurposing or upgrading your legacy assets.

- **Stakeholder mapping**

  Identify all stakeholders involved, including owners, local communities, government bodies and potential customers, to understand their interests and influence.

2 *Strategy development*

- **Sustainability vision and goals setting**

  Define a clear sustainability vision and goals for your legacy assets, based on the initial assessment, which may include energy efficiency, carbon neutrality or waste reduction.

- **Value proposition design**

  Craft a value proposition that leverages the unique characteristics of your legacy assets, such as its historical value, location or existing infrastructure.

- **Business model innovation**

  Develop a business model that aligns with the sustainability goals and value proposition, considering potential revenue streams such as grants, tax incentives or premium services.

- **Create and review the investment case**

  Develop financial models that project costs, savings and revenue streams. Explore different scenarios. Include initial investment, operating costs, expected return on investment (ROI) and payback periods.

3 *Planning and implementation*

- **Feasibility studies and risk assessment**

  Conduct technical and financial feasibility studies to ensure the planned adaptations or business models for your legacy assets are viable. Identify risks associated with the projects, including regulatory risks, market risks and technical risks. Provide strategies for mitigating these risks.

- **Sustainable design and planning**

  Plan for sustainable upgrades, retrofitting or repurposing that respect the asset's legacy while enhancing its value.

- **Collaborative partnerships**

  Build partnerships with technology providers, conservation experts and financial institutions to support the implementation.

4 *Monitoring and adaptation*

- **Performance measurement**

  Establish metrics to monitor the sustainability and financial performance of the legacy asset post-implementation.

- **Continuous improvement**

  Use the performance data to make informed decisions on further improvements or necessary pivots in the business model.

FIGURE 5.1 Process for creating sustainable value propositions and business models for legacy assets

CASE STUDY

Port of Tyne: reinvention through strategic investment and technological innovation

In earlier chapters we saw how the Port of Tyne reoriented its value proposition and business model due to shifting market demands and the imperative of sustainability.

The initial step was to gain credibility by securing new, key customers, which was essential to demonstrate capability without an existing track record in the port's new markets. A crucial part of the strategy was to capitalize on its strategic geographical location. By clearing legacy structures and investing in land remediation, despite the high costs, the port presented itself as a prime location for new investment. Winning early customers played a pivotal role in building trust and demonstrating the port's renewed value proposition.

Then, with the increase in earnings, the port embarked on a critical phase of reinvestment. Mark Stoner, the port's chief financial officer explains:

> The focus was on modernizing equipment which was crucial as the outdated machinery had previously led to operational failures, impacting customer trust. We made significant investments to acquire electric cranes, for example, aligning the port's operations with the decarbonization agenda and future-proofing against evolving environmental standards.

The port also ventured into new territory by focusing on innovation and technology. Understanding the importance of data in the logistics sector, Port of Tyne introduced cutting-edge systems to manage and interpret a wealth of information, aiding in efficient decision-making and operations. This shift in focus from traditional activities to a data-centric approach was instrumental in adapting to the contemporary business landscape.

Mark says:

> The transformation required a solid financial structure. Initially constrained by short-term borrowing, the port transitioned to a more sustainable financing model with long-term facilities and significant capital for investment in game-changing projects. This new financial strategy provided the flexibility to invest in assets and technology with long-term benefits, crucial for the port's growth and future-readiness.

Addressing the broader environmental challenges, the port recognized the need for leadership in green shipping corridors and the selection of future fuels. It acknowledged the importance of circular economy principles, particularly in the context of rare materials and the global shift towards sustainable practices. Profits were reinvested into the business, expanding the port's capacity to borrow and invest further.

Port of Tyne's transformation shows strategic foresight and adaptability. Through leveraging its legacy assets, investing in modernization, embracing technology and securing a stable financial base, the port positioned itself as a leader in sustainable innovation in maritime logistics.

CASE STUDY

Wild Kielder: proactive custodianship of land as a legacy asset

Land is a legacy asset that is important for many organizations engaged in sustainable innovation activities. Ensuring the organization uses and impacts on its land responsibly, while also managing the commercial aspects of its business activities, is important from a corporate sustainability perspective. This becomes particularly important when the area of land is of strong interest to stakeholders.

Forestry England's corporate strategy and biodiversity plan aim to transform the nation's forests into thriving habitats for wildlife in England and build resilience against climate change, while still producing quality commercial timber. The strategy and plan also focus on enhancing community value and contributing to public health and well-being.

The strategy for biodiversity restoration involves re-establishing fully functioning ecosystems through a holistic approach, prioritizing conservation action beyond single species or specific habitats. It emphasizes innovation and collaboration.

The Wild Kielder initiative by Forestry England, launched in 2023, showcases a forward-thinking approach to sustainable innovation. This pioneering forest wilding project is centred on a huge, 6,000-hectare area at Kielderhead in Forestry England's North England region. The area is defined as a place of innovation to rebuild biodiversity by restoring natural processes alongside modern, contemporary forestry.

Wild Kielder aligns environmental objectives with the economic goals of the organization, innovating ways of integrating ecological health with commercial viability and working towards a 100-year vision that prioritizes resilience. Andrew Stringer, head of environment and forest planning at Forestry England, explains: 'Wild Kielder treats the forest as a living legacy asset. The project is adopting a portfolio approach, to diversify conservation methods and build a robust framework capable of withstanding future environmental and economic challenges.'

Tina Taylor, Wild Kielder project officer, adds: 'The scope of the project includes the development of an adaptive forest plan, a detailed monitoring strategy, and an extensive stakeholder engagement strategy, all aimed at improving the forest's biodiversity while maintaining its role as a high-quality timber resource.'

Forestry England is aware that the Wild Kielder project will have an impact on people, nature and the economy, and knows it is important to involve stakeholders in the process. Tina continues:

> The project's success hinges on high levels of stakeholder engagement, where the input from local communities, partners and neighbours is woven into the Wild Kielder vision, and the very fabric of the initiative. Our aim is that this partnership approach enriches the project and ensures that the benefits are shared with the local community.

The project team completed a stakeholder evaluation and analysis internally to identify a diverse range of stakeholders and people with different lived experience and perspectives. They included stakeholders from different industries, including timber, as well as local landowners, parish councils, neighbours, farmers and others.

Stakeholders were invited to a project visioning workshop day at Kielder Forest. The exchange of information and dialogue during exercises enabled the team and their stakeholders to be open and honest about concerns, issues and ideas. They discussed, listened and came up with shared visions together.

Katherine Patterson, North England communications manager, adds:

> Involving stakeholders from the beginning has helped massively to build up enthusiasm and support for the project. Our stakeholders gave us their knowledge and skills to bring this project to life. It has been an incredibly positive journey so far, and most of our stakeholders have become some of our most passionate advocates for the project. Some of them have even offered to 'roll up their sleeves' to help move it forward.

The project is at the early stages of its 100-year journey. However, it has a clear vision and strong stakeholder support that provide a solid foundation for the legacy that will be created for the century ahead.[16]

Buildings and energy efficiency

Buildings use 30 per cent of the world's energy and cause 26 per cent of the energy-related emissions (with 8 per cent from the buildings themselves and 18 per cent from making electricity and heat for them).[17] Countries are getting stricter with rules on how much energy buildings should use, and more buildings are adopting efficient and renewable technology. In 2022, China mandated that all new or updated buildings must be designed to save energy. Japan wants all new buildings to use zero energy by 2030 and existing buildings by 2050. The EU updated its rules in 2023 to require all new public buildings to have zero emissions by 2026, and all new buildings by 2028, and is making the rules for old buildings stricter over time. In the US, the group ASHRAE,[18] an international technical society dedicated to the sustainable advancement of heating, ventilation, air conditioning and refrigeration, set out new 2023 standards for buildings to have net zero energy and carbon use. The aim of the standards is for all new buildings and 20 per cent of old ones to be zero-carbon-ready by 2030, meaning they're super-efficient and use clean energy.[19]

QUICK, SUSTAINABLE INNOVATION WINS FOR LEGACY ASSETS

While sustainably innovating legacy assets can be complex, there are some quick wins that organizations can action in many situations. This checklist may help you to identify options for your sustainable innovation projects.

Energy, water and waste

- **Improving energy efficiency** through retrofitting smart systems for managing electricity, heating and cooling; insulation upgrades; installation of LED lighting for lower energy usage.
- **Integrating renewable energy** can be achieved through solar panels, wind turbines and geothermal heat pumps for sustainable heating and cooling.
- **Water conservation** via rainwater harvesting, greywater recycling and moving taps or faucets to low flow fixtures.
- **Recycling initiatives** for materials such as paper, plastic and metal, and also the dismantling and repurposing of components of out-of-date technology.

Buildings and transport

- **Adaptive reuse of buildings**, i.e. repurposing old buildings for new uses instead of constructing new structures.
- **Green roofs** can be created by converting roof spaces into green areas to improve insulation and reduce runoff.
- **Transportation upgrades** to support electric vehicles, including charging stations.

Technology

- **Transitioning to cloud services** to reduce the need for on-site servers and associated energy costs.
- **Internet of Things (IoT)** usage for monitoring and optimizing the use of resources.
- **Digital documentation** to move to a paperless environment to reduce waste.

Land and outdoor spaces

- **Landscaping** of outdoor spaces for biodiversity.
- **Pollution remediation technologies** to clean contaminated land and water.

- **Increasing green spaces** to enhance human well-being and ecosystem health.

People

- **Workshops and training on sustainable practices** for staff and the community. Options such as the Carbon Literacy Project could be considered, for instance.[20]
- **Offering parts of buildings for community use**, creating a sense of ownership and care for the facilities.

Several of these quick wins are subject to planning permission, the specific requirements of your assets and location, and in many cases the moral, cultural and regulatory obligation to balance the heritage value of legacy assets with modern sustainability requirements, aiming to preserve the past at the same time as embracing the future. While they are not innovations per se, adopting some of these may be innovative for your organization and will have sustainability benefits.

CASE STUDY

Innovative technologies for energy retrofit of a historical building in Venice

From its foundation in 1150 to the present day, the Crucifer Convent in Venice has had a variety of uses. The original purpose set by the Crucifer order for the building was dual: it served as both a place for religious fellowship and a refuge for pilgrims and crusaders en route to the Holy Land. In 1657, the Jesuits took ownership and converted the convent into a school, a role it played until 1773. From the time of Napoleon, the building's southern section was repurposed as military barracks, a function it maintained until 1990. The southern part of the building was then refurbished by the Iuav University of Venice to create a 10,000 m^2 university campus in the heart of a cultural, historic area of the city.[21]

The building was updated to use less energy without ruining its historical character. It was upgraded with new, more efficient heating and cooling systems. A special heat pump that draws from canal water, ventilation that adjusts to how many people are in the building and a combined system for generating electricity and heat, which also helps save energy, were installed. Everything is controlled with a smart system that tracks energy use.

These sustainable innovations cut the building's energy use by 36 per cent without changing its historic integrity and ensured the transformation of a legacy asset for the future.[22]

Sustainable innovation using the principles of circularity

In Chapter 1, circularity, or the circular economy, was defined as a model of production and consumption that involves sharing, reusing, repairing, refurbishing and recycling existing materials and products for as long as possible. This concept contrasts with the traditional, linear economic model of 'take, make, dispose'. We saw how circularity is highly relevant because it encourages the design of projects and systems that minimize waste and make the most of resources. For infrastructure, this means using materials that are durable, repairable and ultimately recyclable at the end of their life. It can also involve innovative approaches such as modular construction, which allows for components to be repurposed or updated rather than needing full-scale replacement.

CIRCULARITY CHECKLIST FOR THE SUSTAINABLE INNOVATION OF LEGACY ASSETS

Adopting circularity in sustainable innovation for legacy technology, buildings, infrastructure and other assets revolves around several key principles and practices. This checklist can help you to find ways to transform your organization's legacy assets into models of circular sustainability, reducing waste and creating more resilient, adaptive systems.

- **Asset audit and life extension**

  Evaluate existing assets for their current state and potential for life extension. Through repair, maintenance and upgrades, explore how you might extend the use-life of products and infrastructure.

- **Material recovery and recycling**

  Dismantle and segregate materials from decommissioned assets for recycling or as inputs for new assets.

- **Repurposing and reuse**

  Consider finding new uses for old assets. For instance, converting disused factories into cultural spaces or apartments, or repurposing old technology for educational purposes.

- **Retrofitting for energy efficiency**

  Create and evaluate plans to update legacy buildings and other assets with modern insulation, smart energy systems, and renewable energy sources to reduce energy consumption and carbon footprint.

- **Modular design**

  When innovating or upgrading, use modular components that can be easily replaced, upgraded or recycled, reducing waste and resource consumption.

- **Sharing platforms**

  Consider developing systems that allow for the sharing or leasing of assets to maximize their use and delay the need for new resources, such as car-sharing services or office space sharing.

- **Digitalization**

  Implement digital tools to monitor and optimize the use and maintenance of assets, ensuring they are used efficiently and maintained effectively.

- **Innovative reconfiguration**

  Adapt infrastructure to serve multiple purposes or to be easily reconfigured for different uses over time.

- **Remanufacturing and refurbishing**

  Instead of discarding outdated technology, remanufacture or refurbish it to meet current standards or new functionalities.

- **Design for disassembly**

  Design new assets with disassembly in mind, so that at the end of their lifecycle, parts can be easily separated and processed appropriately.

- **Assets as a Service (AaaS)**

  Move from selling assets to leasing them, maintaining ownership and responsibility for the asset's lifecycle, and encouraging your supply chain to create sustainable, durable and maintainable products.

- **Local loop**

  Use local materials and resources wherever possible, reducing transportation emissions and supporting local economies.

- **Regenerative resources**

  Use materials that can be regrown or replenished to ensure resources are not depleted.

- **Cross-sector partnerships**

  Collaborate across industries to find innovative uses for waste materials or to share resources efficiently.

- **Policy and incentive alignment**

  Work with governments to create policies and incentives that promote circular practices, such as tax breaks for companies that adopt circular models.

- **Stakeholder engagement**

  Inform and engage stakeholders in the circular process, promoting the adoption of sustainable behaviours and the use of circular services.

CASE STUDY

The Kalundborg Symbiosis: a cross-sector partnership for repurposing waste and resource sharing

In the coastal region of West Zealand, Denmark, the city of Kalundborg has been pioneering the Kalundborg Symbiosis for over half a century. This initiative champions a circular, symbiotic approach to industrial production, where local companies across various sectors collaborate to repurpose waste and share resources such as energy, water and materials. This network of shared resources not only reduces waste but also strengthens the city's economy and environmental resilience. Participating companies include APM Terminals, Argo, Avista Green, Boehringer Ingelheim, Chr. Hansen, COMET, Gyproc, Intertek, Kalundborg Bioenergy, Kalundborg Municipality, Kalundborg Refinery, Kalundborg Utility, Meliora Bio, Novo Nordisk, Novozymes, Unibio and Ørsted.

The vision of the partnership is to be the world's leading industrial symbiosis with a circular approach to production. Its approach has resulted in a robust exchange network with over 30 types of resources circulating among the companies. This collaboration has built a strong foundation of trust and a culture of innovation, benefiting the community, including workers, students and researchers, and significantly cutting down CO_2 emissions annually.

The Kalundborg Symbiosis is recognized as a model for the future, where sustainability and profitability coexist, a standard that they aspire to make widespread. Collectively, the initiative has achieved impressive environmental savings annually, including millions of cubic meters of groundwater, hundreds of thousands of tonnes of CO_2 and tens of thousands of tonnes of recycled materials. Based on the partnership's 2020 published figures, CO_2 emissions within the network have been slashed by 80 per cent since 2015, and the local energy supply has become CO_2-neutral.

The symbiosis supports a significant workforce and involves major companies and local entities working together, illustrating the potential of industrial collaboration in sustainability. The Kalundborg Symbiosis continues to expand its innovative, resource-sharing venture.[23]

CASE STUDY

Microsoft data centres: sustainable, circular asset creation and maintenance

Microsoft Azure is one of the world's largest global cloud data centre providers, alongside organizations such as Amazon Web Services (AWS), Google Cloud Platform (GCP), Oracle Cloud and Alibaba Cloud.[24]

Data centres come in various sizes. They can range from the size of one to two soccer pitches for standard facilities to as large as seven to eight soccer pitches for what's known as a hyperscale data centre. The exact size depends on the data centre's generation and design specifications.

When Microsoft considers global infrastructure and expansion, sustainability is at the heart of its decision-making. The effects on local communities and the natural landscape are paramount.[25,26,27]

Noise pollution is another critical issue. The noise from generators, for instance, affects local wildlife, particularly birds. To mitigate this, Microsoft has opted to encase its backup generators. The organization is also examining additional sound-dampening measures for cooling systems and air handlers.[28]

Construction materials are another consideration. Instead of defaulting to concrete, Microsoft explores sustainable alternatives.[29] Leadership in Energy and Environmental Design (LEED) is a certification system for green and efficient buildings. It is widely used and recognized around the world.[30] Microsoft is committed to all its new data centres achieving LEED Gold status.[31]

Renewable energy and water consumption are also areas of focus.[32] Microsoft aims to secure renewable energy sources and enter long-term contracts to maintain a zero-carbon footprint, achieving what it refers to as '100/100/0', by 2030 – i.e. matching 100 per cent of its electricity consumption 100 per cent of the time, and making 0 carbon energy purchases.[33] Water usage is also critical, especially for data centre cooling. Locations such as Kansas are prioritized for this, where cooling needs are higher than in cooler climates like Sweden.[34]

Microsoft's first data centre in 1989 used traditional methods.[35] Over time, the company has increased server density and optimized space, leading to smaller facilities. Innovations like adiabatic cooling have further enhanced efficiency.[36]

Adiabatic cooling is a process that reduces heat through a change in air pressure caused by volume expansion. In the context of data centres, it involves passing hot, internal air over water-saturated pads, or through a mist. As the water evaporates, it absorbs heat from the air, thereby cooling it. This cooled air is then circulated back into the data centre. This method is especially effective in hot, dry climates, as it uses the outside air's low humidity to achieve cooling without the need for refrigerants or compressors, making it more energy efficient and environmentally friendly than traditional cooling methods.

With each generation of data centres, Microsoft has refined its approach. Its ongoing efforts include AI operations, equipment optimization and striving for the lowest possible power usage effectiveness (PUE). This is a metric used to determine the energy efficiency of a data centre. It is calculated by dividing the total amount of energy used by the data centre by the energy used by the IT equipment alone. The ideal PUE is 1.0, which would mean that all energy used is going directly to computing equipment. However, data centres require additional energy for cooling, lighting and other infrastructure, making a PUE of 1.0 difficult to achieve. The closer the PUE is to 1.0, the more efficient the data centre is. Although a PUE of 1.0 is ideal, it's challenging to achieve due to additional power needs. In 2023, Microsoft's global PUE was 1.18, with the company aiming for 1.12 in future generations, marking continuous progress towards its sustainability goals.[37]

Transparency about Microsoft's energy and water use is part of their commitment to sustainability.[38] The company publishes reports that show energy consumption in a straightforward way, aiming to continuously reduce its PUE.

Microsoft's sustainable innovation strategy for its data centres is proactive, with the intent to evolve and improve, engaging in continuous learning and adaptation.[39] Its earlier facilities were stepping stones, informing the design and operation of subsequent generations through lessons learnt in environmental impact, deployment speed, cost control and capacity.[40]

Conclusion

It's essential to consider legacy assets – technology, buildings, infrastructure and others – as part of your sustainable innovation roadmap, and always bear in mind that the new assets as well as the retrofits that you implement today will be the legacy assets of tomorrow. Use horizon-scanning insights to future-proof your legacy asset innovations as much as is practicably possible. Ensure that each of your sustainable legacy asset innovations has its own value propositions, business model, investment case and forward plan. Pre-empt the tipping point of when the costs of staying the same will outweigh the benefits of making the change and get ready in advance. Consider quick wins as well as more strategic sustainable innovations in combination to achieve your organization's purpose, mission, vision and goals.

ACTION CHECKLIST

- List, review and prioritize your organization's legacy technology, buildings, infrastructure and other assets. Map out any interdependencies between them, and highlight any potentially limiting factors they may create for your other sustainable innovation plans.
- Identify the most appropriate innovation solution for your legacy technology assets, informed by Gartner's seven potential approaches.
- Apply a systems thinking approach to the sustainable innovation of your legacy assets from the outset, being sure not to create overcomplication or overwhelm by focusing on the most important aspects.
- Work through the process for creating sustainable value propositions and business models for legacy assets outlined in this chapter.
- Identify sustainability opportunities for your legacy buildings and infrastructure assets using the quick wins and circularity checklists provided.

Notes

1. Young, D, Remillard, M and Bahl, V. Emissions from legacy assets demand urgent action, BCG, 23 May 2023, www.bcg.com/publications/2023/locked-in-emissions-from-legacy-assets-demand-urgent-action (archived at https://perma.cc/Q2P8-33H5)
2. Plimmer, G. Renewables projects face 10-year wait to connect to electricity grid, Financial Times, 8 May 2022, www.ft.com/content/7c674f56-9028-48a3-8cbf-c1c8b10868ba (archived at https://perma.cc/7UBE-ALWG)
3. Green Switch Capital. How lack of electricity capacity is holding back the green revolution, Green Switch Capital, 6 September 2019, www.gscapital.uk/news/blog/how-lack-of-electricity-capacity-is-holding-back-the-green-revolution (archived at https://perma.cc/T8DA-AQJZ)
4. Akpan, F, Bendiab, G, Shiaeles, S, Karamperidis, S and Michaloliakos, M. Cybersecurity Challenges in the maritime sector, *Network*, 2022, 2 (1), 123–38
5. Irani, Z, Abril, R M, Weerakkody, V, Omar, A and Sivarajah, U. The impact of legacy systems on digital transformation in European public administration: Lesson learned from a multi case analysis, *Government Information Quarterly*, 2023, 40

6 Gartner. 7 options to modernize legacy systems, Gartner, 5 November 2019, www.gartner.com/smarterwithgartner/7-options-to-modernize-legacy-systems (archived at https://perma.cc/MW3G-B2KP)

7 Microsoft (2019) Project Silica proof of concept stores Warner Bros. 'Superman' movie on quartz glass, https://news.microsoft.com/source/features/innovation/ignite-project-silica-superman (archived at https://perma.cc/55SF-YFP6)

8 Microsoft (2022) Project Silica, www.microsoft.com/en-us/research/project/project-silica/ (archived at https://perma.cc/7SZR-MNR2)

9 Interview with the author, 13 November 2023

10 Sellafield Ltd (2023) Our Sellafield: Our people share their stories from 2022/23, https://assets.publishing.service.gov.uk/media/652ff1d792895c000-ddcb9e0/SEL14592_ARoP_ACCESSIBLE_18_10.pdf (archived at https://perma.cc/A49E-EUUQ)

11 Sellafield Ltd (2021) Sellafield Ltd – Procurement routes for innovation, www.gov.uk/government/publications/sellafield-ltd-procurement-routes-for-innovation/sellafield-ltd-procurement-routes-for-innovation (archived at https://perma.cc/U9GV-XGW3)

12 Game Changers (n.d.) About us, www.gamechangers.technology/aboutus (archived at https://perma.cc/SN8F-HZMC)

13 Forrester, J W (2022) *Principles of Systems: Text and Workbook*, Chapters 1–10, System Dynamics Society

14 Schlüter, L, Kørnøv, L, Mortensen, L, Løkke, S, Storrs, K, Lyhne, I and Nors, B. Sustainable business model innovation: Design guidelines for integrating systems thinking principles in tools for early-stage sustainability assessment, *Journal of Cleaner Production*, 2023, 387, 135776

15 Inmarsat (n.d.) Digital now: Why the future of aviation starts with connectivity, www.inmarsat.com/content/dam/inmarsat/corporate/documents/aviation/insights/2022/Inmarsat%20Aviation%20-%20Digital%20Now%20-%20Why%20the%20Future%20of%20Aviation%20Starts%20with%20Connectivity.pdf.gc.pdf (archived at https://perma.cc/5CWX-BM2Z)

16 Interview with the author, 20 February 2024

17 International Energy Agency (n.d.) Buildings, www.iea.org/energy-system/buildings (archived at https://perma.cc/7W79-9FVH)

18 ASHRAE (n.d.) www.ashrae.org/ (archived at https://perma.cc/528Q-WVC6)

19 International Energy Agency (n.d.) Buildings, www.iea.org/energy-system/buildings (archived at https://perma.cc/98NY-HXU2)

20 Carbon Literacy Project (n.d.) https://carbonliteracy.com/ (archived at https://perma.cc/B7GS-BFC6)

21 Università Iuav di Venezia (n.d.) www.iuav.it/INTERNATIO/ABOUT-IUAV/Iuav-profi/index.htm (archived at https://perma.cc/YA6B-GSPG)

22 Schibuola, L, Scarpa, M and Tambani, C. Innovative technologies for energy retrofit of historic buildings: An experimental validation, *Journal of Cultural Heritage*, 2017, 30, 147–54

23 Symbiosis (n.d.) www.symbiosis.dk/en/ (archived at https://perma.cc/KV8N-MSXF)

24 Zhang, M. Top 250 data center companies in the world as of 2025, Dgtl Infra, 14 January 2024, https://dgtlinfra.com/top-data-center-companies/#:~:text=In%20total%2C%20the%2010%20largest,1%2C250%20facilities%20around%20the%20world (archived at https://perma.cc/HJY3-24NJ)

25 Microsoft (n.d.) Datacenter community development overview, https://local.microsoft.com/blog/datacenter-community-development-overview/ (archived at https://perma.cc/PC7C-BLBH)

26 Microsoft. Microsoft announces intent to build a new datacenter region in Finland, accelerating sustainable digital transformation and enabling large scale carbon-free district heating, Microsoft Stories Europe, 17 March 2022, https://news.microsoft.com/europe/2022/03/17/microsoft-announces-intent-to-build-a-new-datacenter-region-in-finland-accelerating-sustainable-digital-transformation-and-enabling-large-scale-carbon-free-district-heating/ (archived at https://perma.cc/7WMV-BHZN)

27 Microsoft (n.d.) Frequently asked questions about our datacenters, https://local.microsoft.com/blog/frequently-asked-questions-about-our-datacenters/#:~:text=Working%20at%20a%20datacenter&text=Typically%2C%20when%20the%20first%20building,time%20and%20vendors%20per%20building (archived at https://perma.cc/T4ZA-NRPX)

28 Microsoft (n.d.) Minimizing noise at our datacenter operations, https://local.microsoft.com/wp-content/uploads/2022/10/Noise-fact-sheet.pdf (archived at https://perma.cc/R9BG-F9RE)

29 Roach, J. Microsoft lays foundation for green building materials of tomorrow' Microsoft Source, 28 September 2023, https://news.microsoft.com/source/features/sustainability/low-carbon-building-materials-for-datacenters/ (archived at https://perma.cc/VX2Y-U7XQ)

30 US Green Building Council (2022) What is LEED certification?, https://support.usgbc.org/hc/en-us/articles/4404406912403-What-is-LEED-certification (archived at https://perma.cc/LF7X-R2FG)

31 Walsh, N. Supporting our customers on the path to net zero: The Microsoft cloud and decarbonization, Official Microsoft Blog, 27 October 2021, https://blogs.microsoft.com/blog/2021/10/27/supporting-our-customers-on-the-path-to-net-zero-the-microsoft-cloud-and-decarbonization/ (archived at https://perma.cc/LR2J-YQ7F)

32 Menear, H. Microsoft to reduce data centre water usage by 94% by 2024, DataCentre, 28 October 2021, https://datacentremagazine.com/critical-environments/microsoft-reduce-data-centre-water-usage-94-2024 (archived at https://perma.cc/MC6X-JW5N)

33 Microsoft (n.d) Powering sustainable transformation, https://datacenters.microsoft.com/globe/powering-sustainable-transformation/ (archived at https://perma.cc/HSQ3-A4FS)

34 Azure (n.d.). Azure global infrastructure, https://azure.microsoft.com/en-gb/explore/global-infrastructure/ (archived at https://perma.cc/W7T5-VBPJ)

35 Microsoft (2015) Microsoft's Cloud Infrastructure, https://download.microsoft.com/download/8/2/9/8297f7c7-ae81-4e99-b1db-d65a01f7a8ef/microsoft_cloud_infrastructure_datacenter_and_network_fact_sheet.pdf (archived at https://perma.cc/4JC9-8DML)

36 Microsoft (2022) Modern datacenter cooling, https://datacenters.microsoft.com/wp-content/uploads/2023/05/Azure_Modern-Datacenter-Cooling_Infographic.pdf (archived at https://perma.cc/B26H-UZEY)

37 Microsoft (2022) 2022 Environmental Sustainability Report, https://query.prod.cms.rt.microsoft.com/cms/api/am/binary/RW15mgm (archived at https://perma.cc/F22N-EUVQ)

38 Walsh, N. How Microsoft measures datacenter water and energy use to improve Azure Cloud sustainability, Azure, 22 April 2022, https://azure.microsoft.com/en-us/blog/how-microsoft-measures-datacenter-water-and-energy-use-to-improve-azure-cloud-sustainability/ (archived at https://perma.cc/M8RD-P53H)

39 Azure (n.d.) Azure sustainability, https://azure.microsoft.com/en-gb/explore/global-infrastructure/sustainability/ (archived at https://perma.cc/QU3S-6SAL)

40 Microsoft (n.d.) Microsoft datacenters, https://datacenters-wp-production.azurewebsites.net/ (archived at https://perma.cc/DG8B-WHEG)

06

Leveraging technology and data for sustainable innovation

CHAPTER OVERVIEW

Chapter 6 delves into how data and technology are pivotal in driving sustainable innovation, especially within complex technical sectors. It provides definitions and sources of different types of data, and offers practical guidance for working with data, including the data skills necessary for an effective sustainable innovation team. The relationship between intuition and data is also discussed, along with the advantages and disadvantages of data use and the importance of having a data strategy for sustainable innovation projects.

The chapter outlines collaborative methods for people and technology, such as the use of big data, digital twins, building information modelling (BIM), cloud-based tools and intelligent asset management. It emphasizes the power of data analytics and digital transformation as a basis for sustainable strategies, applying predictive models to foster growth and efficiency. The chapter explores modern methods for crafting environmental and financial considerations for sustainable initiatives, discussing real-world applications and case studies. It also includes approaches and recommendations for using data and technology to help reduce subjective bias in decision-making.

Introduction

Skilled and capable people, relevant and useable data, and productive and flexible technology are highly valuable and powerful assets for organizations seeking to innovate in pursuit of a significant mission, vision and purpose. And when people, data and technology are used together effectively, the potential for successful sustainable innovation grows. Collaborative human creativity, insights and experience are the spark and drive for new, more sustainable solutions. Data provides context, objectivity and tangibility to guide decision-making. Technology offers processes, systems and connectivity that facilitate data analysis, modelling, monitoring and operational solutions that can transform business operations. To optimize stakeholder value from sustainable innovation programmes and projects, leaders need to consciously find ways to generate synergies by combining the best of what people, data and technology can create together.

In addition, leaders of sustainable innovation need to achieve efficiency and effectiveness within complex settings and requirements. Frequently new assets must be smart, connected, optimized and ready for use within challenging timescales. They also need to comply with safety, environmental and other standards, including new and future ones. Former tools of the trade, such as 2D blueprints and physical documentation, fail to meet current and future-ready criteria by a long stretch. We need to use and develop the new, emerging solutions that are better able to support sustainable innovation in challenging times. Today's goal is to harness the power of centralized, shared data within a cooperative workspace, covering the entire span of an asset's life, from conception to deconstruction and reuse, to extract maximum value from the asset in the most practically and financially sustainable way possible. Innovators are experiencing a transformation in the ways that the planning, design, manufacture, construction and operation of new assets are achieved. Success requires data from all stages and parts of the innovation to inform decision-making. Data-driven, digitally enabled approaches and technologies, combined with the new collaborative delivery models we will cover in Chapters 10, 11 and 12, offer important opportunities and new best practice standards for leaders of sustainable innovation.

What is data?

Definition of data

Data consists of raw elements such as images, words, sentiments, sounds, physical quantities or locations that can be collected, analysed and converted into information and insights to support learning, understanding and decision-making.

Data is unprocessed building blocks – facts and figures without any context. To become useful, data needs to be collated into statistics, i.e. into formats that allow for interpretation, usually to identify trends or patterns. Data becomes information when it has been processed into statistics that provide meaning, context or conclusions. Insights go a step further, offering deeper understanding, implications or actionable conclusions drawn from statistics and information. While data is the starting point, leaders of sustainable innovation need to build capability within the team to be able to convert data into insights to leverage the highest level of knowledge and value from the data and optimize its use as an asset.

Sources of data

Sources of data include, among many others:

- **Numerical** gathered through surveys, experiments and electronic sensors.
- **Text** collected from books, documents, emails and online articles.
- **Image** sourced from digital cameras, satellites and online repositories.
- **Audio** recorded from microphones, phone calls, music and online streaming services.
- **Video** captured by video cameras and smartphones and uploaded on platforms like YouTube.
- **Sensor** obtained from IoT devices, GPS devices and other monitoring systems.

- **Transactional** generated from sales records, financial ledgers and e-commerce platforms.
- **Web** scraped from websites, social media platforms and web analytics tools.
- **Machine-generated** created by computer systems, servers and network logs.
- **Human-generated** derived from user-generated content on social media, feedback forms and crowdsourcing platforms.

Types of data and analytics

There are various types of data. Each type of data and analysis offers a different perspective:

- **Quantitative data** is numerical, allowing for measurable comparisons.
- **Qualitative data** encompasses non-numerical insights, such as feelings, opinions or experiences.
- **Structured data** is organized into a defined format like databases.
- **Unstructured data** lacks a predefined format, like raw text or media.
- **Time series data** comprises observations recorded sequentially over time.
- **Predictive analytics** forecast future events based on historical data.
- **Descriptive analytics** examine past data to understand what happened.
- **Prescriptive analytics** suggest actions to achieve desired outcomes.
- **Diagnostic analytics** delve into data to understand causes of events.
- **Real-time data** is processed at the same time as it's collected, providing immediate insights.

Other data-related definitions

- **Data science** is the multidisciplinary field that deals with extracting knowledge and insights from structured and unstructured data using various processes. It involves the collection, processing, analysis and visualization of data to inform and make decisions. Data science is important for complex sustainable innovation projects because it allows for the analysis and interpretation of complex data sets. By understanding patterns and trends within the data, project teams can make informed decisions, optimize processes and predict future outcomes. This leads to better resource allocation, risk management and measurement of the innovation project's impact on sustainability goals.

- **Artificial intelligence (AI)** includes a wider range of computerized methods and systems that simulate human intelligence. It can include machine learning, where computers learn from data, but also more traditional forms of programmed logic that rely on specific instructions to perform tasks. AI can be hugely beneficial for sustainable innovation projects. It can automate complex tasks, provide intelligent insights and enhance decision-making processes. AI can be used to simulate scenarios and improve efficiency. AI's ability to process vast amounts of data at high speeds helps in rapidly adapting to new information or changes in the sustainable innovation project environment, helping to ensure agility and responsiveness.
- **Machine learning** is a subset of artificial intelligence that involves algorithms and statistical models that computers use to perform tasks without explicit instructions. It focuses on the recognition of patterns and the making of decisions with minimal human intervention, based on the data provided. In the context of sustainable innovation projects, machine learning can predict maintenance needs, optimize energy consumption and reduce waste. It adapts over time, improving the accuracy of forecasts and the efficiency of systems, which is useful for the long-term success and adaptability of sustainable innovation projects.

In its 2023 article 'Tackling climate change with data science and AI', the Alan Turing Institute explained that:

> Data science and AI have a crucial role to play in providing information and tools that will enable the successful development of policies and

interventions for decarbonisation, helping to facilitate change across energy, transport, agricultural and other environment-related systems.[1]

Working with data

Knowing when enough is enough

The idea of having 'perfect' or 'enough' data is unrealistic, owing to the inherent complexities and uncertainties in most sustainable innovation projects. We need to accept that, however much data we have, it is likely to be incomplete, rapidly evolving and subject to biases and inaccuracies. The pursuit of 'perfect' or 'enough' data can lead to 'analysis paralysis', a state where decision-making stalls due to too much emphasis on data collection and analysis. It happens when leaders wait for complete or perfect data before making decisions, which can delay critical actions and slow down project momentum. Sustainable innovation often takes place in a dynamic environment where conditions and requirements can change rapidly. Becoming bogged down in excessive data analysis can result in missed opportunities, outdated strategies and an inability to react to new information or changing circumstances.

Leaders of sustainable innovation need to focus on making informed decisions with the best available data and be prepared to adapt as new data comes in, rather than waiting for perfect certainty that may never arrive. It's important to adopt a flexible and adaptive approach, to continually gather, and analyse, new data and to be prepared to revise strategy as new information emerges, embracing a mindset of continuous learning. For resilient and responsive innovation project management, the key is to make the best decisions possible with the available data, while remaining open to change as and when new or alternative data insights emerge.

Data skills

Data skills are an absolute must for any team engaged in significant sustainable innovation projects, specifically competencies in data literacy, data analysis and data-driven decision-making. Use the data competencies checklist below to complete a skills analysis for your sustainable innovation project team, identify gaps and strengths and create an action plan.

DATA COMPETENCIES CHECKLIST FOR SUSTAINABLE INNOVATION PROJECT TEAMS

Data literacy

- Does each team member demonstrate that they have the required type and level of data literacy for their role?
- Does each team member demonstrate effective critical thinking skills to challenge and question data?
- Does each team member demonstrate that they can communicate data relevant to their role clearly, helpfully and with appropriate impact?

Data analysis

- Do you have a written and communicated data strategy for your sustainable innovation project that clearly identifies the strengths and gaps in your data and how you will collate, analyse and use data at different stages of your project for monitoring, decision-making and continuous improvement?
- Is your data strategy supported by appropriate automation, digitalization and visualization, such as accessible and timely performance dashboards, for example?
- Does your team have someone who is responsible for collating and combining data from across the team, to create a cohesive picture and single version of the truth?
- Does your team use each of the four levels of analytics: descriptive, diagnostic, predictive and prescriptive?

Being data-driven

- Does your team have a culture of using data to inform decision-making, performance monitoring and continuous improvement?

- Does your team action small-scale tests, pilots or experiments that can provide actionable data without the risk of large-scale implementation?
- Does your team combine qualitative data insights from customer or stakeholder feedback with quantitative data?
- Do different functions or team members share data insights openly and transparently in a culture of collaboration?

To dive deeper into how to use data for sustainable innovation projects, and build your team's competence, I strongly recommend the books *Be Data Literate*, *Be Data Analytical*, and *Be Data-Driven* by Jordan Morrow.[2,3,4]

Intuition versus data

Analytical decision-making is often associated with a cognitive preference for structured, data-driven approaches. It involves a methodical analysis of information to arrive at a conclusion. People with this preference tend to rely on logic, facts and a step-by-step process to solve problems. In contrast, intuitive decision-making is more about gut feelings and spontaneous judgements. It's influenced by a person's experiences and emotions, with decisions made quickly based on a holistic view, rather than detailed analysis.

In your team, preferences for analytical and intuitive decision-making may be visible in different ways. Analytical thinkers may prefer detailed reports and metrics, while intuitive members may lean towards idea generation sessions and open discussions. The key to getting the best of both approaches is to create a culture in which both styles are valued and respected, as the reality is that working with data has advantages and disadvantages. Intuition is especially useful for data-informed idea generation, for the exploration of new solutions and for communicating sustainable innovation ideas and proposals to decision-makers and other stakeholders persuasively.

Advantages of working with data

- Effective use of relevant data provides a factual, evidence-based foundation for investment case development and business decision-making. This reduces the risk of costly mistakes in the allocation of resources and the selection of strategies. It can streamline processes and reduce waste and unnecessary cost. This is especially important in sustainable innovation projects where efficiency can lead to environmental benefits.
- Data analytics and modelling, especially predictive analytics, can inform the forecasting of future trends and potential issues before decisions are taken. This allows project leaders to anticipate challenges and opportunities, model a variety of scenarios and position projects for success.
- Data enables the monitoring of a sustainable innovation project's progress against its objectives. By using performance data, leaders can ensure that the project stays on track and delivers the intended sustainability outcomes.
- Data can also be used to communicate the value and progress of sustainable innovation to stakeholders in a clear and quantifiable way, which is essential for maintaining support and funding.
- Many sustainable innovation projects must meet certain regulatory requirements. Data provides the means to ensure compliance and to demonstrate this to regulatory bodies.
- With ongoing data collection and analysis, sustainable innovation projects can evolve through continuous improvement. Data can reveal what's working well and what needs to be adjusted.
- Data can inform the scalability of a project, showing where and how a successful innovation can be replicated in other contexts or geographies, to expand its positive impact.
- In an era where sustainability claims are, quite rightly, scrutinized, data offers transparency and accountability, showing tangible proof of sustainable practices and results.
- Organizations that effectively use data in their sustainability initiatives can achieve a competitive advantage, standing out as innovators and leaders in sustainability.

Disadvantages of working with data and using hypothesis statements

Being data-driven does come with its own set of challenges, however. Of course, the need for data privacy protection and cybersecurity grows as the organization increases its data and becomes more digitally enabled. Integrating new technologies into existing systems presents a technical and cultural challenge. And while data-driven decision-making should, in theory, ensure objectivity, the reality is that all human decision-making, including when it's informed by data, is prone to subjectivity and bias, conscious or otherwise. As we have seen earlier, there is hardly ever such a thing as a 'perfect' data set, and even though there might be access to a super-comprehensive set of data, there is no guarantee that the future will play out as expected.

Humans are always interpreting past data and envisioning the future with our own conscious and unconscious biases. This is at once one of the greatest risks and most interesting opportunities that we have when using data in sustainable innovation projects. To avoid the trap of becoming too complacent and reliant on the data analysis itself, it is essential that leaders of sustainable innovation surface all the significant hypotheses, 'known unknowns' and assumptions that their team and other collaborators are making in the decision process.[5]

A hypothesis is simply 'a supposition or proposed explanation made on the basis of limited evidence as a starting point for further investigation'.[6]

The key features of a well-written, strong hypothesis are that it needs to be:

- An informed view about what your team genuinely thinks will happen
- Written in clear, straightforward language
- Testable

Turning your team's ideas and assumptions into hypothesis statements is a highly effective way of setting up a structure to test them. Where an idea has lots of different components, a separate hypothesis is needed for each. Projects usually succeed or fail based on the assumptions made about demand, deliverability and/or financials, so, for this reason, it's important to include these purposefully and create at least one hypothesis statement for each of the demand, deliverability and financial components of your project.

Assumption surfacing is an effective technique to get your team to identify and share the beliefs and assumptions they have about the potential for your hypothesis to be successful. Team members may then consider if evidence is or could be available to support those assumptions to explore the robustness of the idea. These counter-assumptions should be written alongside each of the original assumptions, and then the potential impact on the sustainable innovation project reviewed in the round.

Sometimes, there is no clear data and answer, and an assumption simply must be made to move forward. In this case, document the assumption you have made, and why. It provides transparency to decision-makers and stakeholders and will help you in future stages of the project. As more information emerges, you may be able to go back and either support your assumption with evidence or change it based on new information, allowing it to become part of your continuous learning process.

When you do need to rely on several key assumptions to move forward, you may find it helpful to model realistic, best-case and worst-case scenarios based on your assumptions in order to forecast a range of potential outcomes, rather than a specific one. You can then take a view on how tolerable your most probable worst case might be for your team and take appropriate action accordingly. We'll explore risk in more detail in the next chapter.

Balancing data and intuition

Leaders of sustainable innovation projects often face the challenge of balancing data-driven decision-making with their own intuition and insights. As we've seen, data-based decisions rely on empirical evidence and analytics. They can reduce subjectivity and bias and help to assess performance, predict trends and make informed forecasts. Intuition-based decision-making leans on experience, gut feeling and heuristic approaches. It's useful when data is scarce or incomplete or when rapid decisions are necessary. Decisions based on intuition can lead to more creative solutions, leading to breakthrough thinking that purely data-driven approaches may miss.

Effective leaders of sustainable innovation get the best of both approaches. They use data to inform and validate their intuitive insights, and conversely sense-check data-driven findings with intuition. Collaborating with a diverse, multidisciplinary team creates a more rounded perspective of complex challenges and decisions as different team members can use their skills and experience to plug data gaps and challenge assumptions, avoid over-reliance on the perspective or bias of a single individual and help take a more balanced view overall.

Data for sustainable innovation

Data and horizon-scanning

In leading sustainable innovation projects, especially in fast-moving, uncertain times, being flexible and responsive is essential. As we saw in Chapter 3, this involves continuous monitoring for emerging challenges and opportunities and analysing the effectiveness of improvised responses. Search, experimentation and strategic agility in formulating and executing new strategic directions are crucial for achieving successful sustainable innovation. The effective use of appropriate data is a critical enabler of spotting the signals of an emerging opportunity or challenge, assessing its potential scale and modelling alternative scenarios and potential responses. Once you have completed your initial horizon scanning from Chapter 3, stay current by developing a plan to maintain your horizon-scanning data through the real-time, or near-real-time, tracking of: market trends; customer behaviours; industry shifts; exchange rates; pricing; competitor and collaborator performance; other political, economic, environmental and social changes.

Data and navigating the innovation process

The sustainable innovation process is one that benefits from the intelligent use of data insights at every stage. It's often depicted as a funnel, shown in Figure 6.1, with the key stages being problem-finding, opportunity recognition and ideation, problem clarification, idea selection and development, and idea testing. The final stage is putting the idea into action, which includes implementation, ongoing operation, maintenance and continuous improvement.

These stages are often shown as a linear sequence, but, in reality, there are multiple movements back and forth between stages. Sustainable innovation should be an iterative process in response to new information and insights that come to light as projects unfold. Innovation requires that we navigate previously unchartered territory. What may seem straightforward at the start may become more challenging later, on hitting unforeseen stumbling blocks and issues. Having the initial idea for the sustainable innovation is usually more straightforward than making that idea happen. Creative problem-solving and judgement, informed by appropriate data, are needed all the way through the innovation process.

FIGURE 6.1 Sustainable innovation process

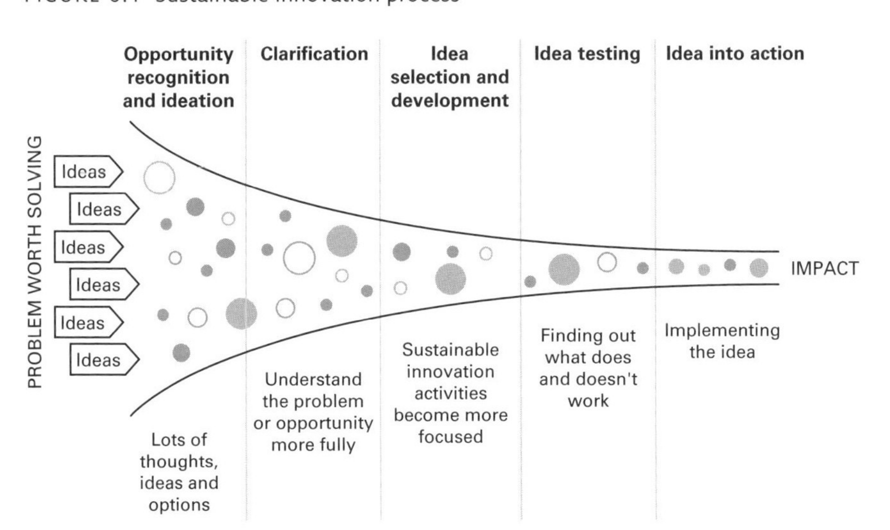

Problem-finding

Problem-finding means identifying stakeholder or organizational challenges that are potentially worth investing time and effort in solving through sustainable innovation. Ideas about these challenges and their relative importance or value can be generated through the horizon-scanning and stakeholder engagement processes we discussed in Chapter 3, as well as through other means.

Opportunity recognition and ideation

In the opportunity recognition and ideation phase of sustainable innovation, data can be the spark that ignites creative thinking. Quantitative data can highlight gaps and inefficiencies, while qualitative data can surface stakeholder concerns and desires. For example, due to China's rapid urban expansion, construction-related noise pollution has emerged as a significant environmental issue. A 2022 study by Wang et al used social media, web crawler technology and text analysis to monitor construction noise pollution through public reactions on these platforms.[7] The resulting data informed ideas for public policy recommendations and potential sustainable innovations.

Clarification

The clarification stage is about understanding the problem or opportunity, and the ideas that you have developed, more fully. It is an important part of the process, as it can avoid unnecessary investment in the later idea development stage for ideas that do not have as much potential as initially thought, or that need to go back in the process before they move forward again. The clarification stage requires a more in-depth evaluation of both the problem and the generated ideas. It's where teams critically assess the viability, relevance and impact of their ideas against the problem they aim to solve. This involves looking at the ideas through various lenses, including feasibility, sustainability and potential for scale.

At this point, it's essential to refine the understanding of the problem or opportunity at hand. This might involve gathering more data, engaging with stakeholders for additional insights or revisiting initial assumptions. The goal is to ensure that the problem is understood as accurately and comprehensively as possible, and that potential solutions will solve the problem in the right way for the organization and its stakeholders.

The clarification stage also involves aligning ideas with the overarching sustainable innovation goals of the project. It's important to determine how well each idea serves the purpose of promoting environmental, social and economic sustainability. This alignment ensures that the ideas contribute positively to broader objectives and are not just sustainable in isolation.

Idea selection and development

As clarity emerges, some ideas will stand out as more promising than others. The clarification stage involves prioritizing these ideas based on their potential impact and strategic fit. The next step in the sustainable innovation funnel is where the most promising ideas are selected and developed, based on this prioritization.

Once an idea is selected, data helps to develop and strengthen it. For example, the Building Innovation Partnership, a collaborative research and innovation enterprise hosted at the University of Canterbury, New Zealand, is committed to enhancing the exchange of infrastructure data across and within organizations. One of its initiatives is to integrate and then share data from different sources. The Building Innovation Partnership integrated Building Information Modelling (BIM), a digital representation of the physical and functional characteristics of a facility, with building asset management data. The aim was to streamline and improve the communication among the various stakeholders involved in a building's entire lifecycle – architects, construction firms and facility or asset managers. The integration process was facilitated by the introduction of client-oriented information supervisors, who are professionals responsible for managing and overseeing the flow of information. They ensured that all parties have access to the most current and relevant data needed for their projects.[8]

In planning sustainable innovation projects, data is indispensable for precision and forecasting. Here, quantitative data can inform resource allocation, timelines and budgeting. Despite Nigeria's suitable climate for renewable solar energy, it hasn't been widely adopted and more than 95 per cent of its energy still comes from fossil fuels. One of the reasons for low levels of renewable energy is that there aren't enough good weather stations that can give the accurate, detailed weather forecasts needed for investors to predict how well solar panels would work now and in the future. To help solve the problem and assist with planning, a machine learning solution was created using 19 years of detailed data from NASA. Although there is more development to be done, the first stages of the modelling of future weather conditions to inform the planning of renewable energy investments look promising.[9]

Idea testing

Idea testing is a critical stage that follows the selection and development of ideas. It involves turning concepts into experiments and tangible prototypes:

- The first step in idea testing is to formulate hypotheses based on the selected ideas. These hypotheses should be specific, testable predictions about how the idea will perform or the impact it will have. It's essential to identify what success looks like for each hypothesis and how it can be measured.
- The next step is to develop prototypes. Prototypes are preliminary models of the sustainable product, service, solution or system that embody the idea in a visual, physical or digital form. They don't have to be perfect but should be sufficient to test the core functions and assumptions of the idea.
- Before a full-scale launch, pilot testing on a small scale can identify unforeseen issues and gather early feedback. The prototype needs to be tested with a sample of real users, or in a representative live

environment, against a robust testing framework to obtain reliable data. This framework should detail the testing methods, the conditions under which tests will be conducted, the participants involved and the metrics for evaluating success.

- Gather data systematically to evaluate the idea against the hypotheses. After the collection phase, a thorough analysis will reveal whether the idea meets the predicted outcomes or if adjustments are needed.
- Ideas rarely pass the testing phase without needing some refinement. Based on the data and feedback, the idea may be modified and retested. This iterative process is fundamental in evolving the idea into a viable solution.
- Testing should also assess the potential risks and broader impacts of the idea. This includes looking at the PESTEL effects we explored in Chapter 3 to ensure that the innovation aligns with the organization's sustainable innovation goals.
- Ensuring that the idea meets regulatory standards is another important aspect of testing. Compliance checks at this stage can save significant time and resources by addressing any legal issues before further development and deployment.
- An idea that works well in a test environment may not necessarily scale effectively. Testing should include assessments of how the idea might perform when scaled up, considering factors such as increased demand, resource constraints and system integration challenges.
- Maintaining detailed records of the testing process, outcomes and lessons learnt is essential. This documentation supports transparency and provides a knowledge base for current and future projects.

Ideas into action

It is essential to plan the launch and implementation of sustainable innovations to maximize their potential for success.

Machine learning algorithms can predict potential failures or identify risks by analysing historical data, allowing organizations to take pre-emptive measures. This can save considerable time and resources by avoiding setbacks and ensuring that the sustainable innovation launches as smoothly as possible.

Every region, ecosystem and community may require a different approach to sustainable innovation. Machine learning algorithms can analyse local data to customize solutions that fit specific needs, increasing the likelihood of successful innovation adoption and maximizing the positive impact on the local environment and population.

During implementation, real-time data allows for agile management and adjustments:

- Advanced analytics can dissect large data sets to determine the most efficient ways to implement sustainable innovations. By understanding which strategies yield the best outcomes, organizations can allocate resources more effectively. For instance, in energy consumption, machine learning can predict peak load times and suggest optimal energy distribution schedules.
- Machine learning models are dynamic and can adapt to new data inputs. This is invaluable in sustainable innovation, where conditions such as market demands, resource availability and environmental factors are in constant flux. As new data is introduced, the models can adjust their predictions and recommendations, ensuring that the sustainable innovation remains relevant and effective.

The Institute of Electrical and Electronics Engineers (IEEE), a professional association, reports that precision agriculture, driven by advancements in technology such as the Internet of Things (IoT), is transforming sustainable farming.[10] Precision farming uses satellite imagery, real-time data from sensors and IoT devices to help farmers monitor and manage crop and soil health more sustainably in real time, with more efficient water use, reduced pesticide needs and overall better crop management. Drones and RFID technology are used for efficient data collection. Advanced analytics and machine learning are also critical tools for processing the vast data and aiding in predictive agriculture.

Ongoing operations and maintenance

Sustainable innovation is not a one-time implementation but a continuous process. Advanced analytics provide ongoing monitoring and feedback, allowing organizations to continuously refine and improve their strategies. This iterative process ensures that sustainable innovations evolve over time and remain as effective as possible. For example, predictive maintenance schedules can be derived from data analytics, helping pre-empt failures and reduce downtime. Data on user behaviour can also influence operational changes for better sustainability practices.

Navigating the 'fog of uncertainty'

The 'fog of uncertainty' is a term that is sometimes used as a metaphor for the unpredictability and the inherent unknowns that often arise during various stages of the sustainable innovation process outlined above.[11] Data can help leaders of sustainable innovation to navigate through this fog, and continue to progress through the funnel, by providing evidence-based insights which can reduce risk and provide predictive analytics to anticipate future trends and challenges.

Data can help penetrate this fog of uncertainty in several ways:

- **Evidence-based insight:** By providing empirical evidence, data can confirm or challenge assumptions, helping to clarify the direction of innovation. It can shed light on user behaviour, market trends and the performance of similar existing solutions.
- **Risk reduction:** Data can identify potential risks and help to assess their likelihood and potential impact. This can inform strategies to mitigate risks early in the innovation process.
- **Understanding of users:** Collecting and analysing data about user preferences, needs and behaviours can lead to a deeper understanding of the market, which can drive user-centred design and innovation.

- **Prototyping and testing:** Data gathered from prototyping, testing and pilot studies can offer concrete feedback on the viability of an idea, informing necessary iterations and refinements.
- **Predictive analysis:** Advanced data analytics can forecast trends and model the potential outcomes of an innovation, enabling proactive decision-making.
- **Decision-making confidence:** With data to back up decisions, innovation teams can move forward with greater confidence, even in the face of some uncertainty.
- **Adaptability and learning:** As data is collected and analysed, innovation teams can adapt their approach based on real-world feedback, turning uncertainty into an opportunity for learning and growth.

CASE STUDY

Creating a data strategy at Port of Tyne

The Port of Tyne, located in North East England, is a significant deep-sea port that plays a crucial role in the UK's maritime industry and regional economy. This port is not just a vital gateway for global trade but also a key contributor to the economy, indirectly supporting over 12,000 regional jobs and contributing more than £621 million to the economy.

The Port of Tyne has introduced a strategic plan called Tyne 2050. This ambitious vision is aligned with the UK government's Maritime 2050 Strategy and the North East Economic Plan. Tyne 2050 is centred on transforming the region and the maritime industry through a series of innovative and sustainable initiatives. The strategy aims to make the Port of Tyne carbon-neutral by 2030, positioning it as a test bed for clean energy and an all-electric port. Additionally, it focuses on doubling diversity by 2030 and leading in technology and innovation, as well as building on the UK's first Maritime Innovation Hub.

The plan represents a commitment to purposeful change and setting industry standards for the future. It emphasizes collaboration, innovation and being an integral part of the North East region. The Port is geared towards enhancing its agility, security and connectivity, ensuring that it continues to thrive and grow, thus supporting regional businesses.

Tyne 2050 is a clear example of how ports can adapt to modern challenges, balancing economic growth with environmental sustainability and social responsibility.

Port of Tyne's Data Strategy is integral to achieving its Tyne 2050 objectives. The strategy's vision is to leverage data as valuable intelligence to enhance safety, efficiency, customer experience, financial sustainability and net zero/ decarbonization goals. It emphasizes democratizing data, ensuring it is user-friendly, accessible and secure, facilitating informed decision-making.

The data strategy includes plans for monitoring KPIs aligned with Tyne 2050 themes and undertaking performance improvement initiatives. A top-down, bottom-up approach is proposed for implementation over two years, focusing on assessing current versus target states, delivering Tyne 2050 projects and conducting intelligence-led performance workshops.

While the Port of Tyne's data strategy focuses internally, its Maritime Data Cluster is about working collaboratively with other ports and maritime organizations around the world. The Maritime Data Cluster is an innovative collaboration between leading international ports, initiated through the UK's 2050 Maritime Innovation Hub, which we'll explore further in Chapter 10. This cluster, the first of its kind, is designed to facilitate the sharing of non-commercial data among ports to address various industry challenges like health and safety, clean energy, decarbonization, cybersecurity and asset management. The Port of Tyne, along with other prominent ports such as ABP, the Bristol Port Company and several others, is a founding member of this cluster.

The Maritime Data Cluster's objectives align closely with the goals of the Port's Tyne 2050 strategy. The collaboration within the Maritime Data Cluster enhances the Port's efforts towards these goals by providing access to a wider pool of data and insights from other leading ports, which can be used to drive innovation and implement sustainable practices.

Furthermore, this initiative supports the Port of Tyne's data strategy. By participating in the Maritime Data Cluster, the Port can leverage shared data to improve decision-making and operational efficiency. This aligns with the data strategy's focus on using data as a valuable tool to enhance safety, efficiency and sustainability. The Maritime Data Cluster provides a platform for the Port to collaborate and share knowledge, which is essential for advancing the technological and environmental aspects of the Tyne 2050 strategy.

The Port uses its private 5G network to test its own sustainable innovations, and those of other organizations, in a live environment, gathering live data for continuous improvement and analysis. Once sustainable innovations have been tested, analysed and are ready for commercial use, they are implemented across the organization. Learning from the sustainable innovation process and appropriate data is shared with industry stakeholders. For example, the Port is a strategic partner in the Connected Places Catapult Maritime Accelerator Challenge. One of the Accelerator projects the Port has tested and supported was ANT Machines' all-electric, cabinless heavy-duty tractor that can pull any trailer or chassis with a fifth wheel. ANT Machines' innovation is designed to replace conventional terminal trucks and make intralogistics more environmentally friendly and efficient.[12]

Data and technology in sustainable innovation projects

Data without people and supporting technology to access, analyse and use it is an unexploited asset. And with the availability of so much data, from such a huge diversity of sources, there is much more volume than even the brainiest of data scientists can be expected to handle successfully. We need technology to help us automate wherever possible the massive task of collecting, storing, processing, modelling and presenting the data to us in ways that we can understand, interrogate and use for practical decision-making. Here we offer the headlines of just a few of the technology approaches and tools for leveraging data for significant sustainable innovation programmes and projects.

Collaborative innovation

With so much data, complexity and diverse stakeholder interests, sustainable innovation project teams, their partners, customers and supply chain need to collaborate more than ever before, as we've seen in the example of the Maritime Data Cluster. We will explore how to achieve successful collaboration later in the book, but at this point it is important to highlight that technology, when used well, creates huge potential for information sharing and collaborative innovation.

Cloud-based collaboration tools break down geographical barriers, allowing for a seamless flow of information and ideas. Real-time data sharing and communication on cloud platforms mean that innovative solutions can be developed and iterated quickly. There are many options for collaborative workspaces for multidisciplinary teams to brainstorm, model and simulate different scenarios for sustainable innovation projects. They enhance transparency, allowing all stakeholders, from designers, architects and engineers to clients and contractors, to access project data. Additionally, these platforms can integrate stakeholder feedback effectively, ensuring that the project evolves in a way that serves the community's needs while also achieving sustainability goals.

Big data

Big data encompasses the vast data sets that are too large and complex for traditional data-processing software but which, when effectively analysed, can reveal patterns, trends and associations. It is often described using the five Vs, said to make up big data's core characteristics: velocity, volume, value, variety and veracity.

The United Nations reports that the quantity of data worldwide is growing at an ever-accelerating rate. In 2020, 64.2 zettabytes of data were created, a 314 per cent increase from 2015. It says: 'A large share of this output is "data exhaust", or passively collected data deriving from everyday interactions with digital products or services, including mobile phones, credit cards, and social media.' The United Nations also states that the proliferation of data is being driven by the widespread collection capabilities of cost-effective and ubiquitous information-sensing mobile devices, along with a global information storage capacity that has been on a doubling trend approximately every 40 months since the 1980s.[13]

For organizations, big data is a strategic asset that, when leveraged with advanced analytics, can provide insights into sustainability practices, enhance operational efficiencies and promote innovation. It allows leaders to make more informed decisions that are based on a balance of evidence and intuition.

Intelligent Asset Management (IAM)

Intelligent Asset Management (IAM) is the integration of digital technologies, such as the Internet of Things (IoT), artificial intelligence (AI), big data analytics and cloud computing, into the management of physical assets. It aims to optimize the performance, efficiency and lifecycle of assets, while minimizing costs and environmental impact.

The key features of IAM are:

- **Real-time monitoring:** Using sensors and IoT devices to continuously monitor asset conditions and performance.
- **Predictive maintenance:** AI and data analytics are used to predict the optimum time for maintenance, preventing breakdowns and extending asset life.
- **Resource optimization and environmental impact:** Optimizing the use of resources such as energy, materials and labour, based on data-driven insights.
- **Lifecycle management:** Managing the entire asset lifecycle from acquisition, operation and maintenance to disposal or recycling.
- **Compliance and reporting:** IAM systems can help organizations meet sustainability regulations and standards by providing accurate data for reporting and compliance purposes.
- **Cost reduction:** While reducing environmental impact, IAM also leads to cost savings by improving efficiency, reducing downtime and extending asset lifespans.

CASE STUDY

Intelligent Asset Management at Northumbrian Water

Northumbrian Water is a private company that provides water services to millions of customers in the North East and South East of England. The company faced a significant challenge when the government transferred ownership of private drains and sewers to local water companies in 2011. This meant approximately 13,500 km of additional wastewater pipes, for which there was very limited information, came under Northumbrian Water's responsibility.

With the transfer of the new network, Northumbrian Water only had about 5 per cent coverage of the necessary data for these assets, leaving a vast amount of the network with unknown attributes such as materials and pipe sizes. Traditional surveying techniques to fill this knowledge gap would have taken an estimated 15 years and cost around £10 million for just the first five years, making it an unsustainable solution.

The company opted for an Intelligent Asset Management (IAM) approach to overcome these challenges. They collaborated to develop a rules-based data inference approach, enabling Northumbrian Water to infer critical data about the wastewater network, such as location, construction and flow direction, based on available information and expert-derived rules.

Their IAM initiative significantly reduced both the time and the costs associated with understanding and integrating the new wastewater assets. The solution helped Northumbrian Water avoid an estimated £8.75 million in costs that would have been incurred through traditional surveying methods.

Northumbrian Water has also been improving their facilities management to capture assets in 3D. This technology has enabled them to accelerate project start-up phases, manage legacy assets more efficiently and reduce site visits, which was especially beneficial during the Covid-19 pandemic. The high-resolution 3D captures are integrated with Northumbrian Water's IAM software, enhancing their operational and engineering teams' ability to manage and maintain assets effectively.

Northumbrian Water's embrace of IAM showcases the power of digital transformation and strategic partnerships in tackling complex infrastructure challenges and setting the stage for future innovation and efficiency in the water management industry.

Digital twins

A digital twin is a dynamic digital representation of a physical asset, process or system. It provides a real-time, holistic view of an asset's performance, maintenance needs and operational efficiency throughout its lifecycle. Digital twins are transformative tools that enable better planning, design, delivery and operation, with the aim to optimize efficiency, minimize waste and enhance safety. By mirroring real-world conditions, digital twins enable scenario planning and problem-solving in a virtual environment, reducing risks and costs associated with physical trials and errors. They rely on big data to accurately reflect the real-world condition of assets and use IAM principles to manage these assets effectively throughout their lifecycle.

Digital twins are typically built using the following key elements: sensors for performance data collection; the data itself, from data and other relevant sources; integration of physical data into the virtual environment; algorithms, models, analytics and visualization; and automated or manual 'actuators', i.e. tools that adjust and make changes as required. They run using several advanced tech tools: IoT; cloud computing; augmented, mixed and virtual reality; and AI.

THE MET OFFICE AND DIGITAL TWINS

The Met Office is the United Kingdom's national weather service, recognized for providing climate-related services. The Met Office UK explains that 'by directly answering asset-centred, user-focussed questions, digital twins represent a transition from hazard-based forecasting into risk-based decision-making'. It says the two features that distinguish digital twins from traditional models are:

- 'A golden thread between the real-world and digital representation.'
- 'A rich user interface that enables non-expert users to interrogate the digital twin.'

The Met Office sees the potential of digital twins in contributing to net zero targets and mitigating the impact of climate change. By integrating AI and machine learning with digital twins, the Met Office can create highly accurate simulations of the environment. These simulations are instrumental for operational planning and can significantly reduce carbon expenditure for data capture.[14]

For leaders of sustainable innovation looking to implement digital twins in projects, it is essential to focus on establishing a robust data infrastructure, ensuring interoperability between systems and maintaining data security and privacy. Collaboration among stakeholders and adherence to industry standards and best practices are crucial for the successful deployment of digital twins.

As infrastructure continues to digitalize, digital twins are increasingly becoming a key element of intelligent asset management capable of dealing with the complexity of sustainable innovation projects.

The Alan Turing Institute asserts that digital twins 'represent one of the most powerful tools' for helping to achieve net zero:

> We envision a wide range of digital twins of both natural and built environments, allowing researchers to optimise systems and policies with the aim of minimising greenhouse gas emissions. For example, a digital twin of an energy system could allow decision makers to explore different policies around renewable energy generation or support schemes for home energy upgrades.[15]

The market for digital twins is growing fast. Research published in early 2023 by FACT.MR shows that it's already worth more than $5 billion globally and is expected to be worth $95 billion by 2033, growing at a compound annual growth rate of 34 per cent.[16]

CASE STUDY

Digital twins at Maersk

Digitalizing supply chain assets, such as warehouses, containers, trucks and vessels, is a huge opportunity for the maritime and logistics industry. Digital twins can enhance supply chain resilience and flexibility. By analysing simulation data, potential bottlenecks, inefficiencies and risks can be identified and mitigated, leading to improved resource allocation and reduced costs. Real-time asset tracking, predictive maintenance and enhanced decision-making are just some of the benefits that digital twins offer. They offer a proactive rather than reactive approach to managing physical assets, helping people to make more informed decisions about resource distribution, planning and risk management, leading to increased operational efficiency and cost optimization.

In early 2023, Maersk published its ambition to create a digital twin of its integrated ecosystem. It's planning to achieve this piece by piece, starting with each individual part of their operation. Once it has all these digital versions, Maersk will connect them all to create a complete digital model of its entire shipping and delivery network. By doing this, Maersk will be able to predict better ways to manage its supply chain processes.

In Maersk's terminals, for instance, digital twin simulations predict operational outcomes in the near term, using data such as container volumes, vessel schedules and terminal activity. This predictive capability helps with precise resource planning, for example the allocation of cranes and ensuring operations are adjusted and optimized in real time.

The scalability of digital twins across different departments within Maersk illustrates the ongoing effort to create a cohesive, interconnected network of twins that links various steps in the supply chain. The company is tackling challenges such as siloed data, and the underutilization of IoT data is also being addressed to optimize processes further.

Maersk sees potential for the integration of digital twins with artificial intelligence and machine learning, achieving wider adoption and interoperability, with great potential for greener logistics solutions, such as supporting a circular maritime economy.[17] Overcoming interoperability challenges and fostering responsible data sharing both within and between organizations are critical to creating resilient, connected and sustainable supply chains. Digital twins need more than technological application; they also require collaboration to achieve a more efficient and environmentally conscious future in logistics.

Maersk also recognizes that human intelligence remains indispensable. Predictive models excel within the range of expected events, but deviations from the norm require human insight to adapt and innovate. The combination of digital twins, predictive AI and human decision-making promises optimal outcomes for Maersk and the wider industry.[18]

Building Information Modelling (BIM)

Building Information Modelling (BIM) is a comprehensive process that involves the generation and management of digital representations of physical and functional characteristics of places. BIM models are files that can be extracted, exchanged or networked to support decision-making regarding a built asset. The approach is used to plan, design, construct, operate and maintain diverse physical infrastructures, such as water, waste, electricity, gas, communication utilities, roads, bridges, ports, tunnels and more.

Both BIM and digital twins are being used more and more for sustainable innovation in construction, enabling smarter project management, reducing waste and improving the efficiency and longevity of buildings. While BIM lays the foundation during the creation phase, digital twins continue to add value, ensuring the asset sustainability and adaptability throughout its lifecycle.

BIM has transformed the management and visualization of building data. It uses cloud-based models to compile multidimensional data, creating an intelligent model from planning to operation and maintenance. Both BIM and digital twins can be used in the architecture, engineering, construction and other industries, for different purposes:

- BIM is a comprehensive process that involves the generation and management of digital representations of the physical and functional characteristics of places. It captures a detailed digital description of every facet of the built asset, playing an important role in the project lifecycle from initial planning to final construction. It allows for advanced 3D modelling and extensive data management and facilitates collaboration between architects, engineers, contractors and other stakeholders, primarily during the planning and construction stages.
- Digital twins provide a dynamic and up-to-date digital reflection of a physical asset. Unlike BIM, their use is not limited to buildings but extends to a diverse range of assets and systems that can be represented in a digital form. Digital twins are useful for monitoring the health and performance of structures, allowing for detailed diagnosis and the simulation of real-world behaviours and conditions. They harness real-time data from IoT devices and sensors, driving predictive analytics to forecast and scenario plan

for future conditions. Digital twins are often leveraged throughout the entire lifespan of an asset or system to support ongoing maintenance and management.

CASE STUDY

Data and technology enabling concurrent design at the European Space Agency

The European Space Agency's Concurrent Design Facility (CDF) has been a pioneer of innovation in the design of space missions, bringing a radical shift from the traditional sequential engineering processes to a collaborative and integrated approach known as concurrent engineering.

Established in 1998, the CDF has fundamentally changed the way spacecraft and missions are designed, allowing for a more efficient and effective process that aligns with sustainable innovation.

At the heart of the CDF's approach is the idea that teams from various engineering disciplines should work together simultaneously, rather than sequentially. This method allows for real-time collaboration and decision-making, which speeds up the design process and leads to more informed choices. The design model at the CDF evolves iteratively, with inputs from different subsystem experts shaping the blueprint as it develops.

Sustainable innovation

Concurrent design (CD) leads to more efficient use of resources. By identifying design conflicts or inefficiencies early, ESA can minimize waste and optimize material use. This efficiency is crucial for sustainable space exploration, where every kilogram of payload counts.

The collaborative nature of CD supports a culture of innovation, where diverse perspectives come together to solve complex problems. This approach has led to breakthroughs in sustainable technology, such as more efficient propulsion systems and satellite designs that reduce space debris.

ESA's CD process includes lifecycle assessments, evaluating the environmental impact of projects from conception to decommissioning. This holistic view encourages the development of sustainable technologies and practices, contributing to the agency's commitment to environmental stewardship.

Data and technology

CD at ESA is an example of how data and technology can revolutionize project management and design processes. By fostering real-time collaboration and leveraging advanced tools, ESA streamlines project timelines and paves the way for sustainable innovation in the space sector.

The CDF is equipped with state-of-the-art data and technology facilities to support this innovative design approach. It houses multiple design rooms with advanced hardware, software and communication tools, creating an environment that facilitates multidisciplinary teamwork. The main design room features a large projection screen and numerous computer stations, while smaller project design rooms support instrument studies and smaller meetings. Additionally, 3D printing facilities allow tangible prototypes to be created, further enhancing the design process.

The CDF's impact extends beyond ESA, with its model inspiring the creation of other design centres across Europe. The facility has played a key role in ESA's Clean Space initiative, which aims to mitigate the environmental impact of space missions. This aligns with the principles of sustainable innovation, as the CDF's flexible environment facilitates the exploration of eco-friendly mission designs.

The CDF's methodology emphasizes customer response, teamwork, trust and sharing, allowing for a consensus-driven decision-making process. This collaborative spirit is not just limited to ESA's internal projects but is also shared with external partners, such as the European Defence Agency, through the provision of consultancy and training services.

By developing a comprehensive understanding of mission design implications, the CDF enables teams to perform various analyses including mission and systems design options, technology gaps identification and systems architectures. This in-depth analysis capability is integral to driving sustainable growth in space exploration.

CD at the ESA represents a forward-thinking approach to spacecraft design and project management. The method, data and technology combined enable multidisciplinary teams to accelerate the design process to drive sustainable innovation. This ensures that space missions of the future are not only technologically advanced but also environmentally considerate.[19,20,21]

Data and technology tips for sustainable innovation

Each organization and sustainable project is different, but here are some tips to help you to draw together the main threads in this chapter and turn them into practical actions that you can implement.

1 Data collection

- **Identify key metrics:** Determine what sustainability and other metrics are most relevant to your project, such as energy consumption, carbon emissions or waste production.
- **Utilize sensors, IoT and other devices:** Deploy sensors, IoT and other devices as appropriate to collect real-time data on resource usage, environmental impact and other factors.

2 Data analysis

- **Implement analytics tools:** Use data analytics platforms to process and analyse the collected data. Look for patterns and trends that indicate areas for improvement.
- **Leverage AI and machine learning:** Employ AI algorithms to predict future trends and optimize operations for sustainability.

3 Benchmarking and goal setting

- **Establish benchmarks:** Compare your data against industry standards or your own organization's historical performance to establish benchmarks.
- **Set sustainable goals:** Based on your data and benchmarks, set specific, measurable sustainability goals and other objectives.

4 Monitoring and reporting

- **Continuous monitoring:** Regularly monitor data to track progress towards your sustainability and other goals.
- **Transparent reporting:** Share your findings and progress with stakeholders through clear, transparent reporting.

5 Making informed decisions

- **Data-driven decisions:** Use the insights gained from your data to make informed decisions about improving sustainability.
- **Iterative improvements:** Continuously refine and adjust your strategies based on ongoing data analysis.

6 Engaging stakeholders

- **Communicate insights:** Share data insights as appropriate with all stakeholders to build awareness and support for sustainability efforts.
- **Collaborative efforts:** Work collaboratively with other organizations, sharing appropriate data and insights to drive industry-wide sustainability.

Other important tips:

- **Start small:** Begin with a manageable scope to effectively implement and learn from data-driven sustainability initiatives.
- **Prioritize accuracy and security:** Ensure the data collected is accurate and secure, maintaining privacy and reliability.
- **Stay informed about the latest technologies:** Keep up to date with the latest advancements in data collection and analysis technologies.
- **Develop a data-competent capability and mindset:** Encourage a team-wide culture that is skilled and motivated to use data insights.

CASE STUDY

The rise of sustainable smart cities through data and technology

As urban areas across the globe tackle growing populations and environmental challenges, data and technology are increasingly being deployed to create more efficient, sustainable and liveable cities.

Smart sensors, an integral component of IoT, are being installed throughout buildings to monitor parameters such as temperature, humidity and occupancy. This data can then be analysed by AI algorithms to efficiently manage heating, ventilation, air conditioning (HVAC) systems and lighting, to reduce energy consumption. For instance, The Edge in Amsterdam, often cited as one of the greenest buildings in the world, leverages IoT for its lighting, using an ethernet-powered LED connected to a central network. Occupants can control the lighting and climate of their individual workspaces via a smartphone app, resulting in energy savings of up to 70 per cent compared to typical office buildings.[22,23]

AI powers predictive models that anticipate traffic congestion and suggest alternative routes to drivers in cities such as Singapore, which employs an Intelligent Transport System that optimizes traffic speeds and reduces the likelihood of traffic jams.[24]

Waste management in smart cities is also seeing a paradigm shift with the adoption of IoT and AI. Smart bins equipped with sensors can monitor waste levels and communicate this information to central waste management systems. AI algorithms then optimize collection routes, reducing fuel consumption and emissions from waste collection vehicles. Barcelona, for instance, has implemented pneumatic waste collection systems in some neighbourhoods, where waste is transported through underground tunnels, significantly reducing the need for waste removal lorries, cutting down emissions and keeping the streets cleaner.[25]

The integration of data and technology into city infrastructures is a current reality, with cities like Amsterdam, Singapore and Barcelona, among others, leading the way. Smart cities leverage the data and connectivity provided by technology to help innovate more sustainable environments where the quality of urban life is significantly improved.

Conclusion

Building the capability and culture to leverage data and technology effectively is becoming even more essential for the design, creation, implementation, operation and maintenance of successful sustainable innovation. It's becoming critical at every stage of the sustainable innovation process. As a result, it's essential that each sustainable innovation programme and project has its own data and technology strategy, and that, where fitting, data and technology collaboration occurs between teams within organizations and externally. At the same time, it is important to be mindful to ensure that privacy, cybersecurity, commercial and sensitive data are managed appropriately. To develop data as an asset, leaders of sustainable innovation need to ensure that their teams have the skills, tools and incentivization to keep data up to date and use it to drive decision-making that optimizes projects and reduces risk, also generating sustainability benefits through increased efficiency. The market and availability for technology to facilitate data usage is growing, and leaders of sustainable innovation will benefit from keeping up to speed with new developments.

ACTION CHECKLIST

- Develop a data strategy for your sustainable innovation programmes and projects, enabling data to become an invaluable asset for your organization.
- Use the data competencies checklist in this chapter to assess the strengths and gaps in your team and develop an action plan to level up their individual and collective competency.
- Be mindful of the advantages and disadvantages of using data in sustainable innovation projects and know when and how to get the best from both data and intuition, noting humans' susceptibility to subjectivity and bias, even when interpreting data.
- Investigate the variety of technology options available to support your sustainable innovations, such as collaborative data platforms, big data, digital twins, intelligent asset management, business information modelling and others.
- Continue to ensure that you have effective measures in place for data privacy, cybersecurity, data maintenance and appropriate protections for commercial and other sensitive data.
- Identify and action useful ways to share data appropriately between teams in your own organization, as well as with external collaborators.

Notes

1 Conner, A, Hosking, S, Lloyd, J, Rao, A, Shaddick, G and Sharan, M (2023) Tackling climate change with data science and AI, *Zenodo*, https://doi.org/10.5281/zenodo.7712968 (archived at https://perma.cc/A3DR-999A)

2 Morrow, J (2023). *Be Data Analytical: How to use analytics to turn data into value*, Kogan Page, London

3 Morrow, J (2021) *Be Data Literate: The data literacy skills everyone needs to succeed*, 1st edn., Kogan Page, London

4 Morrow, J (2022) *Be Data Driven: How organizations can harness the power of data*, 1st edn., Kogan Page, London
5 CNN (2016) RUMSFELD / KNOWNS, YouTube, www.youtube.com/watch?v=REWeBzGuzCc (archived at https://perma.cc/AU3L-P6C2)
6 Dictionary.com (n.d.) Hypothesis, www.dictionary.com/browse/hypothesis (archived at https://perma.cc/KKD2-CEMK)
7 Wang, Y, Wang, G, Li, H, Gong, L and Wu, Z. Mapping and analyzing the construction noise pollution in China using social media platforms, *Environmental Impact Assessment Review*, 2022, 97, 106863
8 Bellamy, L A, Henning, T F P, Amor, R, Jones, D, Pancholy, P, Preston, G and Van Zyl, J E. Data strategies for improving infrastructure value and performance in New Zealand, *Proceedings of the Institution of Civil Engineers – Smart Infrastructure and Construction*, 2023, 176 (3), 98–107, https://doi.org/10.1680/jsmic.22.00008 (archived at https://perma.cc/MCG9-QG8G)
9 Maduabuchi, C, Nsude, C, Eneh, C, Eke, E, Okoli, K, Okpara, E, Idogho, C, Waya, B and Harsito, C. Renewable energy potential estimation using climatic weather-forecasting machine learning algorithms, MDPI, *Energies*, 2023 16 (4), 1603, https://doi.org/10.3390/en16041603 (archived at https://perma.cc/B6DW-UMG5)
10 IEEE Climate Change (n.d.) Tech's role in understanding plant and soil health for sustainable farming, https://climate-change.ieee.org/news/plant-soil-health/ (archived at https://perma.cc/DTV2-RPS3)
11 Santos, J F P and Williamson, P J (2022) Discovering Strategy: Dealing with uncertainty by harnessing serendipity, INSEAD Working Paper, No. 2022/51/STR
12 Connected Places Catapult (2022) Maritime Accelerator Cohort Announced, https://cp.catapult.org.uk/news/maritime-accelerator-cohort-announced/ (archived at https://perma.cc/3KM3-MNUM)
13 United Nations (n.d.) Big Data for Sustainable Development, www.un.org/en/global-issues/big-data-for-sustainable-development (archived at https://perma.cc/PRJ6-AGXP)
14 Met Office (2022) Embedding machine learning and artificial intelligence in weather and climate science and services: A framework for data science in the Met Office (2022–2027), Exeter, UK
15 Conner, A, Hosking, S, Lloyd, J, Rao, A, Shaddick, G and Sharan, M (2023) Tackling climate change with data science and AI, Zenodo, https://doi.org/10.5281/zenodo.7712968 (archived at https://perma.cc/NJ6R-6MK4)
16 Fact.MR (2023) Digital twin market size worth US$ 95 billion by 2033 at of CAGR 34.3%: Fact.MR Research, www.globenewswire.com/news-release/2023/01/18/2591244/0/en/Digital-Twin-Market-Size-worth-US-95-Billion-by-2033-at-of-CAGR-34-3-Fact-MR-Research.html (archived at https://perma.cc/VK2J-SKLZ)

17 Hoffman, H (2023) Four ways digital twins will shake up logistics, Maersk, www.maersk.com/insights/digitalisation/2023/01/31/digital-twin (archived at https://perma.cc/8LP4-R9ZS)

18 Laybourne, G (2023) Transforming decision making in supply chain logistics with digital twins, Maersk, www.maersk.com/news/articles/2023/01/20/transforming-decision-making-in-supply-chain-logistics-with-digital-twins (archived at https://perma.cc/67N5-PDV6)

19 European Space Agency (2019) Concurrent Design Facility, www.esa.int/Enabling_Support/Space_Engineering_Technology/Concurrent_Design_Facility (archived at https://perma.cc/4WG5-GLP7)

20 European Space Agency TEC (n.d.) Concurrent Design Facility, https://technology.esa.int/lab/concurrent-design-facility (archived at https://perma.cc/D55Z-6SLC)

21 European Space Agency (2019) 20 years of ESA's Concurrent Design Facility: an oral history, www.esa.int/Enabling_Support/Space_Engineering_Technology/20_years_of_ESA_s_Concurrent_Design_Facility_an_oral_history (archived at https://perma.cc/DKG6-8J3L)

22 The Edge (n.d.) https://edge.tech/developments/the-edge (archived at https://perma.cc/YNF7-5LG8)

23 Sensoworks (2023) The Edge Amsterdam: Building the future with IoT technology, Sensoworks, www.sensoworks.com/the-edge-amsterdam-building-the-future-with-iot-technology%EF%BF%BC/ (archived at https://perma.cc/93S7-F22N)

24 Smart Nation Singapore (n.d.) Smart Transport Initiatives, www.smartnation.gov.sg/initiatives/transport/ (archived at https://perma.cc/3852-FD95)

25 Haugaard, A (n.d.) Smart Waste management and increasing adoption of IoT solutions, *The Smart City Journal*, www.thesmartcityjournal.com/en/articles/smart-waste-management (archived at https://perma.cc/6LNC-BB67)

07

The sustainable innovation challenge: balancing sustainability, affordability, value, time and risk

CHAPTER OVERVIEW

This chapter examines the complexities of sustainable innovation, where trade-offs and compromises are inevitable to strike a balance between sustainability, affordability, value, time and risk. It addresses the challenge of reconciling these factors, which often seem to conflict, by exploring various strategies to navigate the hurdles in achieving sustainable innovation. It reviews each of the dimensions – sustainability, affordability, value, time and risk – in detail and provides frameworks and models that help to connect and balance all these elements.

The chapter also illustrates the importance of discerning goals from constraints and offers practical advice on how to do this effectively. It distinguishes between genuine constraints and self-imposed limitations and shows how viewing constraints as a creative challenge can be beneficial. The concept of a challenge statement for each innovation initiative is presented as crucial, providing a clear direction for teams to solve problems within real constraints. It highlights the necessity for clear communication and expectation management, especially when constraints significantly impact the ability to meet the objectives of the project.

Finally, Chapter 7 concludes with an overview of a high-level process for managing constraints and risks in sustainable innovation projects, complete with a selection of decision-making techniques that can be applied.

Challenges of sustainable innovation

In most innovation projects there is a tension between achieving all the goals and objectives we have for them and the time, budget, resources and other constraints that we need to work within. There has always been a trade-off between quality and cost, for example, or between customer experience and cost, and this is joined now more than ever by the pressing demands for sustainability, particularly given the climate crisis. Technology is helping to reduce the need for such an either/or approach. The concepts of satisficing and maximizing are both approaches that can be taken. Often sustainable approaches can increase quality and save money, but not always. Every programme and project is different.

CLIMATE CHANGE, AFFORDABILITY, CUSTOMER EXPECTATIONS, SUSTAINABILITY AND INNOVATION IN THE WATER SECTOR

Ofwat, the Water Services Regulation Authority, is the government regulator for the water and sewerage sectors in England and Wales. Ofwat's main role is to regulate the companies that provide these services, with a focus on ensuring that customers receive good quality service at fair prices, while also ensuring that the sector provides for the future and protects the environment. The requirements that Ofwat has set out for water and wastewater companies for their 2025–2030 business plans demonstrate the challenge of addressing multiple needs.

The water sector is grappling with the dual pressures of climate change and the growing expectations of customers, against a backdrop of economic hardship for many. Climate change is leading to drier summers, unpredictable river flows and altered ecosystems. In Wales, future summer droughts may outpace any increase in winter rainfall, while England faces the real possibility of severe drought within three decades.

Customers are demanding better services and are more environmentally conscious, expecting sustainable practices in water use and waste management. An increasing number of households are struggling to afford their bills, so investing in tackling these challenges sustainably and planning for the long term, while maintaining affordability for customers, is essential for companies in the sector.

The response requires a step change in sustainable innovation and collaboration. Companies and governments have set ambitious targets for water conservation, resilience and carbon emissions. To meet these targets, the sector needs to explore nature-based solutions, leverage data and engage with markets to make informed, long-term decisions. A transformative approach is necessary.[1,2,3]

Just a few examples of how the UK water sector is delivering sustainable innovation to meet these challenges include:

- Smart water metering by Anglian Water, Thames Water and Northumbrian Water. This offers key insights into how customers use water through almost instant data. It helps them make smart choices, spots leaks early, both in the network and at homes, and consequently supports sustainability.[4]
- Developing solar AI systems for use at water pumping stations to adjust power use across the network as needed.[5]
- Severn Trent Water is testing the conversion of plastic and wet wipes found in wastewater to energy. These wipes are a significant pollutant affecting animals and wildlife, as well as being a major cause of sewer blockages.[6]

By 2023, Ofwat's £200 million Innovation Fund had awarded more than £105 million to projects across England and Wales that were 'pioneering solutions to improve the water sector's environmental impact – from the transition to net zero to tackling leaks and spills – and to transform services for consumers'. Since they began in 2020, Ofwat's innovation challenges have led water companies to form new cross-sector sustainable innovation collaborations with gas and electric utility providers, universities, charities, small businesses, tech giants, local governments and community organizations. Previous funding winners included 'robots that patrol water and wastewater pipes to pinpoint cracks; carbon capture technology to convert CO_2 into household products; seagrass meadow restoration off the Welsh coast, and projects to power city buses with gas extracted from sewage'.[7]

The challenge of apparently mutually exclusive criteria

Sustainable innovation projects face a range of factors and some of them may be at odds with each other. For smaller, less-complex projects, an intuitive or more traditional approach can be taken, but, as we've seen in Chapter 6, this will not work for substantial ones. In larger, technical projects, where some of the project criteria appear to be mutually exclusive, there will invariably be differences of opinion between stakeholders and decision-makers over what should be a priority.

Satisficing and maximizing

When the ambition is to create something truly ground-breaking and ambitious, the risk is that too many compromises are made, leading to a solution that is dissatisfying and mediocre. What can happen, especially in corporate and some public sector sustainable innovation projects, is that the desire to please many people, and to tick the right boxes, leads to 'satisficing' instead of 'maximizing'. Satisficing is an adaptive strategy. It means settling for a solution that, in the round, is good enough.[8,9] Maximizing, on the other hand, means aiming for the very best solution possible in the circumstances.

Sometimes, satisficing is indeed the best approach to take, for example when time or information are limited and a decision simply must be made, or when the potential gain of working on an even better solution is marginal and not worthwhile. Every situation is different. In general, the more complex the project, the greater the number of key stakeholders involved; and the tighter the budget and time constraints, the more likely it is that satisficing will be a practical and appropriate approach.

CASE STUDY

The Spittelau waste-to-energy plant, Vienna: successful satisficing

The Spittelau waste-to-energy plant in Vienna, Austria, is an example of the principle of satisficing in sustainable innovation projects. Originally built in the 1960s, the facility underwent significant revamping and optimization in the early 1990s, after a major fire in 1987. The facade was given its current distinctive and colourful appearance by the artist Friedensreich Hundertwasser.[10] The architectural design and external artwork of the Spittelau waste-to-energy facility is a hallmark feature of the plant, contributing not only to Vienna's skyline but also to its identity as a city that values sustainability and art.

The distinctive design and artwork reflect Vienna's commitment to integrating sustainability with cultural and artistic expression. The renovation of the facility was a technical and environmental upgrade as well as an opportunity to showcase how industrial infrastructure can contribute positively to urban aesthetics and community engagement.

Hundertwasser's design is characterized by undulating shapes, vibrant colours and the incorporation of natural elements, such as green roofs. This approach to an industrial facility's design has made the plant a tourist attraction and a source of local pride, illustrating that functional infrastructure can also enhance the urban environment's visual and cultural aspects.

Since 2009, the site has been home to a district cooling centre, which Wien Energie uses to provide environmentally friendly cooling.

Later, in 2013, further updates were signed off for the Spittelau. This time the aim was to improve waste-management practices and diversify fuel and electricity supply in Austria. With a total cost of approximately €144 million and European Investment Bank finance of €70 million, the project's objectives were ambitious, targeting a reduction in CO_2 emissions and positioning the site as a highly efficient combined heat and power plant.[11]

In 2022, a new power-to-heat plant to convert surplus electricity from the grid into heat was opened at the Spittelau. This has helped to stabilize the power grid and developed the site to network the electricity and heat sectors more closely to move away from fossil fuels.[12]

The integration of state-of-the-art control technology allowed for virtualization and simulation that not only enhanced operational efficiency but also provided a hardware-independent solution. This minimized maintenance costs and used available technological advancements to meet project requirements effectively, without pushing the boundaries of innovation to the maximum.

The Spittelau facility's continuous operation for over 50 years and its supply of district heating to Viennese households show its success as a sustainable infrastructure asset. The use of the latest controls technology ensures that the plant remains a significant contributor to waste disposal and resource utilization in the city.

The Spittelau's developments demonstrate the principle of satisficing by achieving a satisfactory level of performance that meets environmental, economic and social goals without necessarily maximizing every potential aspect of the project. It also reflects the pragmatic trade-offs that are often necessary in large-scale sustainable innovation projects, where the ideal of maximizing performance must be balanced against real-world constraints and regulatory requirements.[13,14,15]

How to prepare for the best solution you can achieve within your constraints

Separate your goals from your constraints

Once you've defined your goals, as discussed in Chapter 2, the next key step to achieving the optimal solution for your sustainable innovation is to specify your constraints. Be very clear about the difference and relationship between the two. When goals and constraints become intertwined and are seen as inseparable, clarity of thought and decision-making become more difficult.

As a reminder, goals are the desired results or outcomes that a project aims to achieve. Constraints, on the other hand, are the parameters and requirements that contain and shape the sustainable innovation. They can be financial (such as budget limits), temporal (deadlines), technical (available technology), regulatory (laws and standards) or resource-based (materials and human resources). Constraints must be identified, understood and managed to successfully navigate the project towards its goals.

The relationship between goals and constraints is often a dynamic interplay where the constraints define the boundaries within which the goals must be pursued. For example, a goal may be to build a zero-emissions building, but budgetary constraints may limit the types of materials and technologies that can be used, leading to a compromise that still advances sustainability but may not achieve zero emissions. Another example might be a goal to construct a waste-to-energy facility for decarbonization that faces resistance from the local community who live near the proposed site.

Instead of viewing constraints as limitations, try to treat them as a helpful framework that grounds the specific context and application of your sustainable innovation project. There are numerous benefits of having clearly defined and understood constraints:

- Constraints can help to narrow down the scope of a project, ensuring that teams are not overwhelmed by too many options or directions. They force teams to focus on what is most important and feasible within the limits of the project.
- When constraints are present, project teams are pushed to think creatively and look for alternative solutions. This can lead to more innovative approaches that might not have been considered if there were no limitations. As designer and architect Charles Eames said, 'design depends largely on constraints'. The most accomplished innovators frame constraints as a puzzle to be solved.
- Constraints related to resources, whether financial, material or human, encourage the efficient use of available assets. This efficiency is at the heart of sustainability, of course, which aims to do more with less and minimize waste.
- By defining what cannot be done or what is non-negotiable, constraints help in identifying potential risks early on. This allows teams to devise strategies to mitigate these risks before they become problems. Instead of addressing each constraint in isolation, seek ways to integrate them into the project's process or design. For instance, regulatory constraints can be woven into the project timeline from the start.
- Constraints force project managers and teams to prioritize certain goals over others, which can lead to a more strategic approach to project planning and execution.
- Constraints ensure that project goals are realistic and achievable. They prevent the pursuit of unattainable ideals that can lead to project failure.
- Regulatory and legal constraints ensure that the project complies with relevant laws and standards, which is crucial for the project's acceptance by stakeholders and the market.
- Constraints can provide clear criteria against which the project's success can be measured. If a project overcomes its constraints to meet its objectives, it is easier to quantify its achievements.

Tips for verifying genuine constraints

It is essential to make sure that your project constraints are genuine and not merely assumed. There is a difference between a real constraint and a self-imposed limitation that is unnecessarily restrictive. Examples of assumed constraints include thinking that: senior managers will or will not support your ideas before you've spoken to them; innovative solutions may be too costly or time-consuming before you've explored them; it's not possible to find a better solution than the one used today. If you're in doubt, use the hypothesis testing and assumption surfacing processes described in Chapter 6 to challenge your thinking. Adopt a methodical and analytical approach, combined with a willingness to explore new possibilities.

Additional tips for differentiating between genuine constraints and self-imposed limitations are:

- **Look for evidence.** Genuine constraints are often backed by data, research or historical evidence. Leaders should look for empirical data that supports the constraint.
- **Get expert advice.** Consulting with experts or stakeholders can provide insight into whether a limitation is an industry standard or regulatory requirement (genuine) or a perceived barrier that may be more flexible.
- **Benchmarking.** Comparing the project to similar successful projects can inform you whether certain limitations are commonly encountered and navigated (genuine) or if they have been overcome elsewhere.
- **Complete a risk analysis.** Conducting a risk analysis can help determine the impact of constraints. Genuine constraints often have significant risks associated with them, while self-imposed limitations may be based on risk-averse attitudes that can be mitigated.

CASE STUDY

Clear identification of goals and constraints: Open Nuclear Platform's innovation challenge statements

The National Nuclear Laboratory (NNL) initiated the Open Nuclear Platform to address technical challenges in the nuclear sector. This platform invites external organizations to present innovative solutions, with the potential to secure funding and partnership support.[16]

NNL is a nuclear services technology provider that is owned and operated by the government and that covers the whole of the nuclear fuel cycle. A leader in nuclear innovation, NNL supports the UK's nuclear industry and aims to tackle global issues across health and nuclear medicine, clean energy, environmental restoration and security. By launching Open Nuclear, NNL seeks to cultivate new partnerships and utilize external expertise.

Organizations with relevant innovative proposals can apply for up to £12,000 in 'Discover' project funding to validate the viability and impact of their solutions. After the 'Discover' phase, there are further development opportunities in collaboration with NNL.

Technical challenges are clearly articulated and published on the platform as Challenge Statement documents. These provide an overview and context for the innovation challenge, an opportunity statement which outlines the challenge goals, followed by a summary of the challenge constraints and next steps.

An example challenge from 2023 is for the remote characterization of mixed contaminated waste.[17] The goals in the opportunity section of this challenge are described as follows:

> NNL is seeking technology which can be used to characterise and classify the Red Bunker waste pile to identify the waste objects contained within the bunker and their levels of radioactivity... Characterising the waste pile would give a long term forecast of the processing time for this volume, which currently cannot be provided. It is anticipated that successful delivery of this project could serve as an example of how to adopt a systematic approach to waste clearance, leading to increased regulator confidence... Ideally, NNL would like a 3D map of the whole waste pile which gives material profiles. The waste held in Red Bunker has built up over time and is of mixed materials. There are significant unknowns regarding the composition of the waste and the levels of activity, as there are no detailed records of what was added to the bunker over many years of use... As a minimum, it would be helpful to understand where the metal items are within the waste pile as this

> would inform the approach for clean-up and provide intelligent estimates of what is there and timescales for removal.

The document also describes the constraints:

> The waste pile is surrounded by cast concrete blocks which are around 0.5 m^3. The concrete blocks form the walls for Red Bunker on two sides and a concrete wall forms the third side. There is no roof or covering directly on top of the waste pile.
>
> The pile can be accessed via an Electric Overhead Travelling (EOT) crane which can carry up to 15 tonnes.
>
> The whole waste pile and concrete blocks sit on top of a concrete floor. The concrete blocks which contain it form an area of around 2m wide by 5m long by 2m high. It is possible to be within approximately 3m of the waste pile without having to wear the additional PPE currently used for processing.
>
> Access to the pile is via stairs at either end of the building and cranage access is through a 3m by 3m hatch. An antechamber trolley runs from the hatch to the crane to move items in and out, and the trolley can take items between 2m and 5m in length.
>
> All sides of the waste pile are accessible for characterisation which could be done from above or around the top of the blocks.
>
> Large waste items at the bottom of the pile are more difficult to deal with so an accurate picture of what is inside the pile will help plan their removal.
>
> If shielded access is required into the bunker, then the operator is not physically constrained, but time would be constrained and would need to be carefully monitored, with likely less than one hour of access time possible.

Through clear, meticulously detailed articulation of the project goals, and the constraints of the challenge, NNL has created a specific, bounded framework for innovators to problem-solve within.

Five dimensions of sustainable innovation projects

There are five dimensions that apply to most sustainable innovation projects. These are value, sustainability, affordability, time and risk. These dimensions can and do often overlap. They are not always mutually exclusive – for instance, more time can also mean spending more money.

Within sustainable innovation projects, your goal is to achieve the specific value sustainable proposition that you are seeking to create for your various stakeholders, as we've explored in Chapters 4 and 5.

However, there may be some other elements of sustainability that are constraints that lead to further innovation.

CASE STUDY

Chengdu Science Fiction Museum: value and sustainability

The Chengdu Science Fiction Museum, with its design reminiscent of a scene from a science fiction narrative, is an example of both innovation and sustainability in the heart of Sichuan's capital. The museum's futuristic design, created by Zaha Hadid Architects, harmonizes with the local geography, reflecting mountainous silhouettes on one side while mirroring cloud formations on the other. This dynamic structure, appearing to float upon the lake, changes perspective with the viewer's angle, a deliberate feature to ensure a continuously evolving visual experience.

This project's original vision encompassed both an architectural wonder and a sustainable marvel. The museum's rapid construction, completed in only 12 months – a process that typically spans several years – was a feat achieved through the extensive use of digital design tools. These tools allowed for a harmonious design and construction process, ensuring that all stakeholders, from architects to builders, worked with collective precision and understanding.

Sustainability elements from the original vision are evident in the museum's integration with the local climate. The building's expansive roof, which hosts solar panels, is designed with specific angles to provide shade and minimize the reliance on artificial cooling systems. The use of natural light through skylights and extensive windows further reduces energy consumption during daylight hours.

Constraints played a pivotal role in shaping the project's sustainability profile. Local material sourcing was a necessity not just for expedience but also to reduce the environmental impact associated with transportation. A modular aluminium panel system was developed for the roof, striking a balance between the desired seamless aesthetic and the need for rapid, cost-effective production.

The Chengdu Science Fiction Museum's role extends beyond a mere exhibition space. It is a central piece in the city's 'Future City' development in the Pidu district, symbolizing Chengdu's growing emphasis on science and technology. The museum houses science fiction artefacts and embodies the city's aspiration to be a hub for innovation.

Typically, a project of this scale would take around five years to complete, but the timeline was dramatically shortened to just one year to meet this deadline of hosting the World Science Fiction Convention in October 2023. Advanced software played a crucial role in meeting the tight deadline. It enabled simultaneous design and construction processes, allowing for rapid progress without sacrificing quality or safety.[18]

By marrying the original sustainability goals with the practical constraints imposed by the project's ambitious timeline and environmental considerations, the Chengdu Science Fiction Museum stands as a superb example of sustainable architectural achievement in modern China.[19]

Create a challenge statement for each of your sustainable innovation projects

As early as possible in your sustainable innovation process, create a challenge statement for your project. This statement should clearly articulate the project's goals and objectives and detail all the specific, material constraints, including budgets, timescales, regulatory requirements and quality standards, to the best of your knowledge at the time. Circulate this challenge statement to key decision-makers and stakeholders to get and document feedback and alignment on your project from the outset. Spending time doing this upfront will be a good investment, as it will save time on clarification and sorting misunderstandings later. However, don't try to please and involve too many people. Identify those individuals with high interest in and high influence over your project and focus on them first. These are the decision-makers and influencers who can make or break your plans.

'How might we...?' and 'What if...?' questions

To accomplish the most valuable sustainable innovation outcomes in line with your project goals, while also achieving the affordability, time, risk and any other parameters important to your project, your leadership mindset and the mindset you nurture across your team and supply chain are of the utmost importance. While perfect solutions rarely, if ever, exist, a creative problem-solving mindset that perseveres with asking 'how might we...?' and 'what if...?' rather than defaulting to finding reasons why things can't be done is beneficial.

Working with challenging constraints requires a shift from linear, analytical thinking to a more dynamic, divergent way of exploring problems. The 'how might we...?' question framework is designed to open up the problem space and invite innovative thinking, transforming challenges into opportunities. For example, 'How might we use renewable energy sources to power this facility despite budget constraints?' By framing the issue as a question, it becomes an invitation to explore a range of solutions rather than being a roadblock. Asking 'what if...?' encourages thinking beyond the current constraints and exploring wider possibilities. For example, 'What if we could achieve the same result with different, more sustainable materials?' This question can lead to the discovery of alternative methods or materials that hadn't been considered before.

Avoid simply sitting around a meeting table and discussing the challenges. Use approaches such as visual thinking with sketches, diagrams, flowcharts and mind maps to conceptualize problems and solutions. This helps to simplify complex concepts, making it easier to communicate ideas and see connections between different elements of a problem. For example, your team could create storyboards to visually map out user experiences or project timelines, or infographics to represent data or processes visually. Alternatively, systems thinking, discussed in Chapter 5, helps to map out how all parts of the project are interconnected, and how changes in one part can affect the whole.

The use of 3D models and prototypes allows teams to explore solutions in a tangible form. This can be especially useful in sustainable innovation projects where the interaction of physical materials and design is crucial. Technologies such as 3D printing can quickly create models of proposed solutions, allowing for hands-on experimentation and testing, helping teams to iterate and refine ideas more effectively.

Take a look at Chapter 11 for a deeper dive into ideation and knowledge exchange tools, techniques and approaches for creative problem-solving with multidisciplinary teams.

What happens if your constraints mean that your vision and goals are genuinely unachievable?

Even with the most can-do mindset and highly talented lateral thinkers on board, sometimes the constraints mean that the team is unable to create a solution that achieves the goals of its sustainable innovation project. Unfortunately, what happens all too often is that either stakeholders and decision-makers refuse to change their expectations and requirements and/or the project team persist in their commitment to the pursuit of the initial objective, even though they know achieving it is highly improbable.

Of course, organizations, teams and individuals have different levels of optimism or pessimism about what can be achieved, as well as varying tolerances for and interpretation of risk. People also have different levels of confidence when it comes to challenging what is being asked of them and pushing back appropriately. Equally, leaders can vary in their individual receptiveness to appropriate challenge and expectation management.

Whether your sustainable innovation project goals appear to be feasible or not, to navigate the complexities of human expectations, perceptions and concerns, it's important to use a structured process to work through how achievable those goals are given the constraints and risks, and what might be done to mitigate those risks. Do this as early as possible in your sustainable innovation process and keep repeating it at regular intervals until your project is complete to incorporate the new insights and learning your team acquires as the project progresses. Working through this as a team, and involving external experts and key supply chain partners, helps to bring balance by bringing diverse perspectives to your explorations and decision-making processes. Remember, too, that sustainable innovation is an iterative journey, not a straight line. You're working to achieve something that hasn't been done before. Be prepared to be flexible, learn and adapt along the way and ensure that any differences in expectation versus reality are resolved at every stage of project development and delivery.

Process for managing constraints and risks in sustainable innovation projects

The process for navigating the achievement of goals within the constraints of complex and challenging sustainable innovation projects, while also surfacing and mitigating key risks, requires a series of steps that involve analysis, collaborative problem-solving, and continuous adaptation. Here are the steps of my Sustainable Innovation Constraint and Risk Management Process. This process can be used at different stages of sustainable innovation project delivery.

SUSTAINABLE INNOVATION CONSTRAINT AND RISK MANAGEMENT PROCESS

Step 1: Goal-constraint mapping

- **Objective:** Establish a clear understanding of project goals and constraints.
- **Activities:**
  - List project goals and constraints.
  - Map constraints to goals to see where they align or conflict.
  - Consider constraints based on their impact and rigidity.

Step 2: Sustainability integration review

- **Objective:** Ensure sustainability is integrated into all aspects of the project.
- **Activities:**
  - Review project plans and prototypes for sustainability alignment.
  - Adjust project plans to better meet sustainability goals.

Step 3: Constraint impact analysis

- **Objective:** Analyse the impact of each constraint on project goals.
- **Activities:**
  - Perform impact and probability analysis for each constraint using a scale (e.g. high, medium, low).
  - Determine potential bottlenecks and showstoppers.

Step 4: Risk identification

- **Objective:** Identify and categorize risks associated with constraints.
- **Activities:**
  - Carry out a SWOT (strengths, weaknesses, opportunities and threats) analysis to identify risks.
  - Categorize risks into groups (e.g. technical, regulatory, financial).
  - Give each risk a rating for probability and impact using a scale (e.g. high, medium, low).

Step 5: Constraint and risk mitigation strategies

- **Objective:** Develop strategies to mitigate identified constraints and risks.
- **Activities:**
  - Brainstorm mitigation strategies for high-impact, high-probability constraints and risks.
  - Create a mitigation plan for each, with clear actions and responsibilities.

This process is designed to be cyclical, with the insights from later stages feeding back into earlier ones. It should promote a dynamic and responsive approach to managing constraints and risks in sustainable innovation projects.

Decision-making tools and models

Several decision-making tools and models can support teams in finding the right balance between value, sustainability, affordability, time and risk. Using two or more of these tools in combination can provide a comprehensive approach to decision-making, ensuring that the five dimensions are considered and balanced effectively throughout the project's development and implementation.

MULTI-CRITERIA DECISION ANALYSIS (MCDA)

Multi-criteria decision analysis (MCDA) helps in evaluating different options based on several criteria. In this case, this would be the five dimensions of value, sustainability, affordability, time and risk. It allows for a structured and systematic comparison of various project alternatives and their impacts and is a useful tool for informing sustainable innovation projects, where decision-makers often face a complex interplay of factors that must be balanced to reach a satisfactory outcome.

1. The MCDA process begins with a clear definition of the decision context and the identification of the various criteria that will influence the final choice. These criteria may encompass a range of sustainability indicators, economic considerations, technological feasibilities, social impacts and environmental benefits.
2. Once the criteria are established, the next step is to collect and analyse relevant data for each criterion, ensuring an evidence-based approach to the decision-making process. MCDA usually requires assigning weights to each criterion, reflecting their relative importance in the context of the project's goals. This step is critical and often involves stakeholder engagement to align the weighting with the values and priorities of those who will be affected by the decision.
3. With the criteria weighted, the MCDA process then moves to score each option against the criteria. Scoring can be quantitative or qualitative, but it should be consistent across all options to ensure comparability.
4. The compilation of weighted scores leads to an overall ranking of options, providing a transparent rationale for why one option may be preferred over another. MCDA can also include a sensitivity analysis to understand how changes in criteria weights or scores

affect the outcome. This sensitivity analysis can highlight which criteria are the most influential and help to identify areas where further information could change the decision.

5. The final phase of MCDA is the review and interpretation of results in consultation with all key stakeholders. This collaborative review is vital to validate the decision analysis and to secure buy-in for the chosen approach. By carefully considering each criterion and involving stakeholders in the process, MCDA helps ensure that decisions for sustainable innovation projects are robust, equitable and aligned with broader organizational and sustainable goals.[20,21]

The Triple Bottom Line framework

The Triple Bottom Line (TBL) framework expands the traditional reporting framework to consider environmental and social performance in addition to financial performance.[22,23] To use the TBL method for decision-making analysis in sustainable innovation projects, follow these practical steps:

1. Define the three pillars:
   - Economic viability: Assess the financial aspects of the innovation. Will it be profitable? How does it impact the organization's bottom line, or profitability?
   - Environmental sustainability: Evaluate the environmental impact. How does the innovation affect the environment? Does it use resources more efficiently?
   - Social equity: Consider the social implications. Does the innovation contribute to the community, for example, or improve employee well-being?
2. Set objectives for each pillar:
   - Develop clear, measurable goals for each of the three pillars that align with the overall objectives of the project.
3. Identify stakeholders:
   - Identify who will be affected by the sustainable innovation project and understand their interests and concerns.

4. Gather data:
    - Collect data relevant to each of the TBL's three pillars. This could include financial projections, environmental impact assessments and social impact studies.
5. Develop indicators:
    - Create indicators that can measure performance in each area – for example, profit margins for economic viability, carbon footprint for environmental sustainability and job creation for social equity.
6. Weight each pillar:
    - Depending on the project and the organization's goals and values, assign weights to each pillar to reflect their relative importance.
7. Analyse trade-offs:
    - Evaluate the trade-offs between the pillars. For example, a more expensive process that is environmentally friendly may be justified by the social and environmental benefits it provides.
8. Make decisions:
    - Use the data and analysis to make informed decisions that balance all three pillars. Ensure that decisions align with the long-term strategy of the organization and its commitment to creating all the benefits detailed in its value propositions for different stakeholders.
9. Monitor and measure:
    - After implementation, continually monitor the outcomes and measure against the TBL objectives. Adjust the project as necessary to maintain balance among the three pillars.
10. Report:
    - Transparently report the outcomes to stakeholders, showing how the project performs against each pillar of the TBL.

LIFECYCLE ASSESSMENT

Lifecycle assessment is used to assess the environmental impacts associated with all the stages of an innovation asset's life. This can inform decisions that balance the five dimensions of sustainable innovation projects and consider the broader implications of the project.

Additionally, it can help balance short-term with long-term requirements and benefits. These stages include raw material extraction, manufacturing, distribution, use and disposal or recycling.[24,25,26]

To use lifecycle assessment in practice, begin by defining the scope of the analysis, including the boundaries of the life stages to be assessed. Next, collect data on the inputs and outputs of each stage – for example, energy and material usage, emissions and waste production. This data can then be used to evaluate the environmental impacts, such as carbon footprint, water usage and pollution, along with other factors.

Using the results from the analysis, identify areas where the innovation can be improved to reduce negative environmental impacts, save time or money and generate more value. This might involve choosing different materials, altering the production process or designing the product for easier recycling at the end of its life.

When balancing the five dimensions of sustainable innovation – economic, environmental, social, technical and temporal – this method can provide a detailed picture of the sustainability impacts, which can be weighed against economic costs, technical feasibility and project timelines. This holistic view ensures that decisions are made with an understanding of sustainability that spans the entire lifecycle of the sustainable innovation, rather than focusing on a single aspect or short-term gains.

REAL OPTIONS ANALYSIS

Real options analysis (ROA) is a financial modelling technique that helps evaluate and manage the uncertainty and flexibility inherent in sustainable innovation projects. It provides a framework for quantifying the value of potential opportunities and the cost of risks associated with them, offering a way to adaptively steer projects in the face of changing conditions.[27,28,29]

There are eight key steps to implementing ROA in sustainable innovation projects:

1. Identify the real options: Begin by pinpointing opportunities within the project where you have the flexibility to make decisions in the

future – options to expand, delay, modify or abandon the project. These options are real because they represent real-world, actionable choices.

2 Assess the project lifecycle: Map out the timeline of the project and the points at which options might be exercised in the future. Sustainable innovation often involves long-term processes, so it's important to consider how external factors, such as market volatility or regulatory changes, might evolve.
3 Estimate cash flows: Forecast the expected cash flows for the sustainable innovation project, both with and without the exercise of the options. These cash flows should reflect the costs and revenues associated with the project's value propositions and business model.
4 Determine volatility: Assess the volatility of the sustainable innovation's returns, which is a measure of risk. In sustainable innovation projects, volatility might come from technological change, market demand for sustainable products or environmental regulation shifts, for example.
5 Model the options to assess financial and non-financial value, using the data you've collated in steps 1–4 above.
6 Build in flexibility for decision-making ROA emphasizes the value of flexibility. When new information is available, or circumstances change, update the analysis to reflect these changes. Keeping the analysis updated helps with making decisions that keep the sustainable innovation project aligned with its goals and vision, while managing economic and other risks.
7 Execute the options: When it is strategically favourable to do so, execute the preferred real options identified. This could mean scaling up operations or pivoting away from less sustainable practices as better technologies become available, for example.
8 Monitor and review: Sustainable innovation projects are dynamic. Regularly review the project's progress and the external environment. Update the ROA accordingly to ensure decisions continue to balance risks and opportunities effectively.

Conclusion

In sustainable innovation projects, challenges and constraints are inevitable. Instead of viewing these as obstacles, they can be seen as catalysts for creative thinking and innovative solutions. Maximizing stakeholder value often involves a trade-off between the ideal and the achievable, always aiming to optimize the project's value in alignment with the organization's vision.

Leaders of sustainable innovation will often have to make decisions on satisficing versus maximizing to balance diverse stakeholder priorities with the practicalities of making progress on the project.

Clarity is key – it's crucial to distinguish between overarching goals and specific objectives. It's equally important to scrutinize whether constraints are truly external or if they stem from the team's own limitations or assumptions. Crafting a clear and comprehensive challenge statement at the start is beneficial. This statement should encapsulate the project's vision, objectives and the true constraints, serving as a touchstone to scope the project, align expectations and document intentions.

While a proactive and resourceful attitude is essential in devising solutions that meet the project's aims within real limitations, it's equally important to be realistic about what can be promised. Sustainable innovation leaders should identify and communicate potential issues early, managing expectations and planning for contingencies.

Laying out constraints and risks methodically, along with strategies to mitigate them, helps prevent overcommitment and excessive caution. This process supports decision-makers by providing solid reasons why a project is considered viable and outlines the proposed approach to managing it.

Utilizing decision-making tools as discussed can aid in objectively evaluating the practicality, likelihood of success and appeal of different methods, ensuring that choices made are grounded in reality and enhance the sustainable innovation project's prospects for success.

ACTION CHECKLIST

- Consider and document how your sustainable innovation may need to satisfice a number of different requirements and create a challenge statement for the project as early in the process as you can. Keep updating this as circumstances change or as new information comes to light. Separate your goals from your constraints.
- Test that constraints are genuine, using hypothesis testing and assumption surfacing methods.
- Develop a culture of lateral thinking and problem-solving in your team, framing goals and constraints as a creative challenge and as being helpful for the project, rather than as barriers that prevent progress and innovation.
- That said, be ambitious and realistic about what can be achieved given the constraints. If the targeted value from the sustainable innovation is genuinely unachievable, this needs to be raised and addressed with key decision-makers as soon as possible so that alternative plans can be agreed.
- To inform and design decisions and action plans, use the process for managing constraints and risks in sustainable innovation projects, along with an appropriate selection of decision-making tools and models presented in this chapter.

Notes

1 Ofwat (n.d.) PR24 and beyond: Creating tomorrow, together, www.ofwat.gov.uk/regulated-companies/price-review/2024-price-review/framework-and-methodology/pr24-and-beyond-creating-tomorrow-together/ (archived at https://perma.cc/EN92-6NB9)

2 Addleshaw Goddard (2021) Ofwat consults on the 2024 Water Sector Price Review (PR24), www.addleshawgoddard.com/en/insights/insights-briefings/2021/energy/ofwat-consults-on-the-2024-water-sector-price-review-pr24/ (archived at https://perma.cc/7BCD-BJZ5)

3 PA (n.d.) PR24: Shaping and preparing for the price control, www.paconsulting.com/industries/energy-and-utilities/pr24-shaping-and-preparing-for-the-price-control (archived at https://perma.cc/KR6F-W6WE)

4 Welsh, A. Innovation is the key to solving the UK's water challenges, Xylem, 13 October 2023, www.xylem.com/en-uk/support/lets-solve-water-blog/innovation-is-the-key-to-solving-the-uks-water-challenges/ (archived at https://perma.cc/V6JP-K8RA)

5 Largue, P. Power generation drives sustainability in UK's water sector, Power Engineering International, 4 July 2023, www.powerengineeringint.com/renewables/strategic-development/power-generation-innovation-drives-sustainability-in-uks-water-sector/ (archived at https://perma.cc/6X43-ZUV6)

6 Hannis, M. Why innovation is crucial to successful, sustainable water management, Innovation News Network, 2 January 2024, www.ofwat.gov.uk/ofwats-innovation-fund-seeks-bold-and-ambitious-entries-to-water-breakthrough-challenge-4/ (archived at https://perma.cc/LH4G-Z9AF)

7 Ofwat (2023) Ofwat's Innovation Fund seeks bold and ambitious entries to Water Breakthrough Challenge 4, www.ofwat.gov.uk/ofwats-innovation-fund-seeks-bold-and-ambitious-entries-to-water-breakthrough-challenge-4/ (archived at https://perma.cc/CB87-GWE6)

8 Simon, H A. Rational Choice and the Structure of the Environment, *Psychological Review*, 1956, 63 (2), 129–38

9 Simon, H A (1957) *Models of Man, Social and Rational: Mathematical essays on rational human behavior in a social setting*, Wiley, New York

10 Mathieson, J (2023) Building of the month: September 2023 – Spittelau Incinerator, Vienna', Twentieth Century Society, https://c20society.org.uk/building-of-the-month/spittelau-incinerator-vienna (archived at https://perma.cc/5BXB-V9DK)

11 European Investment Bank (2023) Wien Energie Wate to Energy, www.eib.org/en/projects/all/20120235 (archived at https://perma.cc/P6FH-B9C3)

12 KLINGER Holding (2023) Industrial valves provide 'Power-2-Heat' for district heating, www.klinger-international.com/en/news/industrial-valves-provide-power-2-heat-for-district-heating (archived at https://perma.cc/57PJ-QDB8)

13 Siemens Energy (n.d.) Wien Energie, Austria: State-of-the-art control system for simulation and hardware independence, www.siemens-energy.com/global/en/home/stories/waste-to-energy-plant-wien-energie.html (archived at https://perma.cc/96B5-D4Y6)

14 European Investment Bank (2023) Wien Energie Wate to Energy, www.eib.org/en/projects/all/20120235 (archived at https://perma.cc/2KDU-76N9)

15 Wien Energie (n.d.) Müllverbrennungsanlage Spittelau, www.wienenergie.at/privat/erleben/standorte/muellverwertungs-anlage-spittelau/ (archived at https://perma.cc/4GCE-ZDYA)

16 National Nuclear Laboratory (n.d.) Open Nuclear Platform, www.nnl.co.uk/innovation-science-and-technology/showreel/collaborations/open-nuclear-platform/ (archived at https://perma.cc/L585-7XDA)

17 National Nuclear Laboratory (2023) Challenge statement: Remote characterisation of mixed contaminated waste challenge, https://wazoku-clients.s3.eu-west-2.amazonaws.com/nnlinnovation.wazoku.com/Documents/ON_892_Red_Bunker_Challenge_Remote_Characterisation_of_Mixed_Contaminated_Waste.pdf (archived at https://perma.cc/DUN6-5B6Q)

18 Kenyon, N. Zaha Hadid Architects' latest project is real-life science fiction, Boss Hunting, 2 November 2023, www.bosshunting.com.au/lifestyle/design/chengdu-science-fiction-museum/ (archived at https://perma.cc/NMA3-YWB6)

19 Cairns, R. This massive 'floating' museum is straight out of science fiction, CNN Style, 29 October 2023, https://edition.cnn.com/2023/10/30/style/chengdu-science-fiction-museum-hnk-spc-intl/index.html (archived at https://perma.cc/SG57-T4WG)

20 Kitsios, F and Kamariotou, M. Information systems strategy and innovation: Analyzing perceptions using multiple criteria decision analysis, *IEEE Transactions on Engineering Management*, 2023, 70 (5), 1977–85, DOI: 10.1109/TEM.2021.3103318, https://ieeexplore.ieee.org/document/9535154 (archived at https://perma.cc/LX4L-AVMW)

21 Nalmpantis, D, Roukouni, A, Genitsaris, E, Stamelou, A and Naniopoulos, A. Evaluation of innovative ideas for Public Transport proposed by citizens using Multi-Criteria Decision Analysis (MCDA), *European Transport Research Review*, 2019, 11 (22), https://etrr.springeropen.com/articles/10.1186/s12544-019-0356-6 (archived at https://perma.cc/GJY6-CDZR)

22 Miller, K. The TRIPLE BOTTOM LINE: What is it & why it's important, Harvard Business School Online, 8 December 2020, https://online.hbs.edu/blog/post/what-is-the-triple-bottom-line (archived at https://perma.cc/R27E-XZAY)

23 Open University (n.d.) The triple bottom line, Sustainable innovations in enterprises, www.open.edu/openlearn/money-business/sustainable-innovations-enterprises/content-section-3.2.3 (archived at https://perma.cc/3QNK-YUXU)

24 Life Cycle Initiative (n.d.) Life Cycle Sustainability Assessment, www.lifecycleinitiative.org/starting-life-cycle-thinking/life-cycle-approaches/life-cycle-sustainability-assessment/ (archived at https://perma.cc/M3J2-R65M)

25 Barbieri, R and Santos, D F L. Sustainable business models and eco-innovation: A life cycle assessment, *Journal of Cleaner Production*, 2020, 266, 121954, https://doi.org/10.1016/j.jclepro.2020.121954 (archived at https://perma.cc/E92P-9QFA)

26 Open University (n.d) The life cycle assessment (LCA), Sustainable Innovations in enterprises, www.open.edu/openlearn/money-business/sustainable-innovations-enterprises/content-section-3.2.1 (archived at https://perma.cc/LU3K-JVPJ)

27 Faster Capital (2024) Innovation: Fostering innovation with Real Options Analysis, https://fastercapital.com/content/Innovation--Fostering-Innovation-with-Real-Options-Analysis.html (archived at https://perma.cc/2D8E-TMSP)

28 Marks, M, Ryder, D, Liu, J and Krans, P. An economic approach to investing in climate adaptation, *ICF*, 22 September 2021, www.icf.com/insights/environment/real-options-analysis-climate-resilience-investment (archived at https://perma.cc/XQ5B-KJD8)

29 Stroombergen, A and Lawrence, J. A novel illustration of real options analysis to address the problem of probabilities under deep uncertainty and changing climate risk, *Climate Risk Management*, 2022, 38, 100458, https://doi.org/10.1016/j.crm.2022.100458 (archived at https://perma.cc/49TF-R3LM)

08

Sustainable innovation decision-making and governance

CHAPTER OVERVIEW

Sustainable innovation project failure often stems from weak project initiation, ineffective leadership or insufficient sponsorship. Governance plays a key role in these areas and is crucial for turning things around. Effective governance facilitates a culture where leaders can thrive. It combines empowered individuals, efficient structures and clear reporting for timely decision-making. Trust among team members is crucial. Relying solely on contracts is a weak replacement for a cohesive and effective leadership team.

Good governance is transparent, encourages challenges, aids in sound decision-making and recognizes outcomes and lifetime value. While technical expertise is essential, it's just the basic expectation from a supplier. In sustainable innovation projects, governance needs to be more responsive, more agile and more efficient to achieve timely decision-making, escalation and success.

Chapter 8 is dedicated to exploring why an updated approach to governance is needed for sustainable innovation, providing insights and strategies for effective approaches within this context. It proposes strategies for effective governance in sustainable innovation, and for measuring success via a sustainable innovation KPI dashboard. The chapter also features options for a sustainable innovation governance structure and programme sponsorship model.

The role of governance in sustainable innovation

As we've seen in Chapter 7, decision-making for sustainable innovation projects can be complex and challenging. Where the stakes are high, it's essential that sustainable innovation projects are supported by effective governance and sponsorship. Corporate governance is a multifaceted concept involving a set of rules, practices and processes by which an organization is directed and controlled. It balances the interests of an organization's various stakeholders, such as shareholders, management, customers, suppliers, financiers and the community. The principles of strong corporate governance include accountability, transparency, fairness and responsibility, which are vital for fostering internal cooperation, reassuring shareholders and promoting the company's image to its stakeholders and the public.[1,2]

Corporate governance provides the structure and principles necessary to guide and support sustainable innovation projects. The governance of these projects, in turn, reflects and reinforces the organization's commitment to ethical, sustainable and innovative practices. It's important to integrate sustainable innovation into the core governance framework for the following reasons:

- Effective corporate governance can drive the success of sustainable innovation projects, while the learnings from these projects can enhance governance practices.
- Corporate governance frameworks demand accountability and transparency, which are crucial for the governance of sustainable innovation projects. These principles ensure that projects are executed ethically and report their impact accurately.
- Governance structures must adapt to the unique challenges and opportunities of sustainable innovation. This includes embracing new technologies, responding to environmental challenges and meeting evolving stakeholder expectations.
- The governance of sustainable innovation projects involves ethical considerations, such as environmental stewardship and social responsibility. Corporate governance plays a critical role in embedding these values into the organization's operations.

Corporate governance and ESG

In Chapter 1, ESG was defined as focusing on three specific areas: environmental, social and governance factors, often used by investors and financial analysts to evaluate a company's performance. It typically involves publishing ESG metrics and performance in sustainability reports, often for investors and regulatory compliance. ESG is often driven by the need to meet consumer and investor expectations, reduce risks and enhance financial performance through ESG integration.

ESG is a fundamental component of corporate governance today and is increasingly incorporated through shareholder oversight. Shareholders, responsible for the governance of a corporation, ensure that companies not only maintain a sound management structure and demonstrate ethical practices but also consider their environmental and social impact. This shift is driven by a growing demand from investors, who seek to align their investments with their values. By integrating ESG into governance frameworks, corporations can better assess and mitigate risks associated with these factors.[3]

Studies have indicated a positive correlation between strong ESG performance and overall corporate performance. Research by PwC in 2021 showed that 79 per cent of the 325 global investors surveyed consider ESG risks and opportunities an important factor in investment decision-making, and only 33 per cent said they believe the quality of current ESG reporting, on average, is good.[4] ESG reporting, driven by regulatory compliance and consumer demand, has become much more of a priority. This is especially true for large companies, which face greater expectations and scrutiny from consumers and shareholders. Consequently, effective governance of ESG issues can lead to better risk management, improved brand reputation and increased shareholder value. The board of directors is instrumental in integrating ESG into the organization's overall business strategy. It is responsible for translating broad ESG aims into concrete strategies and for corporate performance and reporting. The board's involvement ensures that ESG goals are not only set but also effectively implemented across the organization.

Incorporating ESG into corporate governance is not without its challenges. Companies face the task of prioritizing different ESG aspects based on their industry, geographical location and material risks. For instance, companies in heavy industry might focus more on environmental aspects, while those in healthcare could prioritize social issues. The complexity of each ESG pillar requires companies to be selective and strategic in addressing the demands and expectations of various stakeholders. In addition, the fast rate of change and increasing global complexity continue to make ESG oversight challenging.

Governance and greenwashing

Greenwashing is the practice of making false or misleading claims about the environmental benefits of a product, service or a company's policies to appear more environmentally friendly. It is used by companies, deliberately or unintentionally, to appeal to environmentally conscious consumers without making the levels of environmental improvements claimed.

Greenwashing is a governance issue because it involves misleading stakeholders about a company's environmental practices or impacts. This can lead to a breach of trust and accountability, which are core elements of corporate governance. When organizations engage in greenwashing, they undermine the principles of transparency and honesty that are essential for good governance. Greenwashing can also result in a lack of progress in addressing environmental challenges, as it allows organizations to appear more environmentally responsible than they are, leading to fewer changes and improvements.

There are also various laws and regulations that address greenwashing, primarily through consumer protection and advertising standards. These laws require truthful and non-misleading advertising, with specific regulations addressing environmental claims. For example, the Federal Trade Commission (FTC) in the United States has guidelines to ensure that marketing claims about the environmental attributes of products are truthful and not misleading. The guidelines cover a variety of topics, including general environmental benefit claims, certification and seals of approval, compostable claims, degradable claims, free-of claims, non-toxic claims and claims about products made with renewable energy or materials.[5] Effective corporate and sustainable innovation project governance prevents greenwashing by ensuring transparency and accuracy in environmental claims. They establish clear policies and guidelines for environmental marketing, conduct regular audits of environmental claims and build a culture of honesty and responsibility within the organization. Additionally, they engage stakeholders in open communication about the company's environmental practices and performance.

MAERSK'S ESG GOVERNANCE

In its 2023 Sustainability Report,[6] A.P. Moller – Maersk (Maersk) set out its extensive ESG (environmental, social and governance) strategy with accelerated targets to achieve net zero greenhouse gas emissions by 2040. This strategy positions ESG at the core of the company's purpose and operations.

Maersk has been dedicated to sustainable practices for over 10 years, and published its first ESG strategy in 2022. The company's ESG strategy is integral to its business, key to its success and a way to deliver value to customers. It tackles significant sustainability issues, risks and opportunities for Maersk and revolves around three core commitments:

- Environment: 'We will take leadership in the decarbonisation of logistics.'
- Social: 'We will ensure that our people thrive at work by providing a safe and inspiring workplace.'
- Governance: 'We operate based on responsible business practices.'

These commitments are areas where Maersk aims to excel and lead globally, leveraging its unique skills, resources and strengths, for example in decarbonization, where it seeks to set new industry standards and address customer and societal challenges. Each commitment has corresponding ESG topics, KPIs (key performance indicators) and targets:

- Environment: climate change, environment, ecosystems.

- Social: safety and security, diversity, equity and inclusion, human capital, employee relations and labour rights, human rights, sustainable/inclusive trade.
- Governance: business ethics, data ethics, sustainable procurement, responsible tax, citizenship, governance.

The KPIs and metrics are used for internal tracking and external reporting. Recognizing the dynamic nature of ESG issues, Maersk plans to continue evolving its priorities, target-setting and reporting.

In 2023, Maersk introduced a dedicated ESG Committee within its Board to guide and shape the company's ESG strategy. This committee offers valuable feedback to the Executive Leadership Team and equips the Board with detailed strategy insights on ESG issues. The committee convenes every three months to engage in in-depth discussions on key strategic ESG topics.

The strategy is aligned with the UN Sustainable Development Goals, ESG benchmarks, science-based targets and stakeholder expectations.

Each of the 14 ESG themes is led by a member of the Executive Leadership Team (ELT), responsible for driving initiatives and delivering intended outcomes. The governance model includes various committees and advisory groups for different ESG aspects.

The ESG Steering Committee, a central committee presided over by an ELT member, oversees ESG governance, performance management and external reporting.

Maersk's governance framework, called Commit, integrates ESG topics related to compliance and responsible business practices. It ensures internal governance through the Maersk Values, Code of Conduct and Maersk Rules. The company also emphasizes responsible inorganic growth, incorporating ESG risk assessments into its mergers and acquisitions process, to align acquisitions with its ESG strategy.

SOURCE Maersk 2023 Sustainability Report

Sustainable innovation project governance

Sustainable innovation project governance integrates ESG factors into the project-management process. This approach ensures that sustainability is considered in decision-making, aiming to balance economic viability with environmental and social responsibility. It is the framework that guides how decisions are made for projects within an organization. It includes a set of activities and guidelines for planning, executing and managing projects. The main components of project governance include the structure of the team, the people involved and the information flow within the project. It focuses on ensuring that projects are run smoothly, stay on budget, meet their objectives and deliver the target value proposition.

Corporate and sustainable innovation project governance

The relationship between project governance and corporate governance lies in their shared goal of ensuring accountability and effective management, but they differ in their focus and scope. Corporate governance is the broader system by which organizations are directed and controlled, balancing the interests of different stakeholders.

In contrast, sustainable innovation project governance specifically addresses the management and oversight of individual programmes and projects within an organization. It provides a decision-making framework for projects, ensuring they align with the company's strategic objectives and are efficiently executed. Sustainable innovation project governance is a subset of corporate governance, dealing with a more focused area of management.

Sustainable project governance operates under three main pillars: structure, which involves the organization's support and vision for the project; people, emphasizing the importance of effective project managers; and information, which is critical for clear and consistent communication throughout the project. We'll explore these pillars in more detail in this chapter.

Effective governance of sustainable innovation programmes and projects

A programme is a collection of interconnected sustainable innovation projects. Effective governance gives confidence to people developing sustainable innovation programmes and projects, those approving them and those investing in them. The way an organization is governed overall, and at individual sustainable innovation programme and project level, impacts its ability to innovate successfully. Significant sustainable innovation projects often encounter problems in their early stages, and suboptimal governance constrains project delivery and limits value creation from the investment as explained by the UK's Infrastructure and Projects Authority, 2023:

> Having good governance will enable key decisions to be made with confidence throughout the project lifecycle. It will make sure that there is a 'single controlling mind' for the project and will establish the right levels of accountability and authority. This will help to keep the outcomes aligned with strategic objectives, manage the risks, and realise the planned benefits.[7]

Ensuring maximum value from sustainable innovation projects

The collective role of those participating in the governance of sustainable innovation projects is to:

- Set strategic direction through the programme, ensuring alignment with corporate objectives.
- Oversee implementation of sustainable innovation project and operating plans, ensuring costs, risks and asset system performance are controlled across the asset lifecycle phases, including approving investment plan proposals.
- Oversee the sustainable innovation risk management framework, ensuring relevant risks are identified, assessed and managed.
- Ensure compliance with applicable legislation, regulatory and statutory obligations.
- Monitor performance of the sustainable innovation in delivering corporate objectives, including service, asset and plan delivery performance.
- Actively seek opportunities and prioritize actions for continuous improvement, embracing good practice from other industries and sectors.

- Be accountable for establishing, documenting, implementing, maintaining and continually improving the organization's sustainable innovation and asset management system.
- Agree the course of action where performance changes and/or change in regulations impact overall operational and capital expenditure and innovation performance.
- Ensure integrated planning across projects.

Common issues affecting successful delivery of sustainable innovation projects

In complex sustainable innovation projects, several challenges often arise, each with the potential to negatively impact successful delivery and value creation. Leaders of sustainable innovation may find that their programmes will benefit from reviewing their organization's governance approach, to identify and resolve any issues that could be leading to sub-optimal performance. The most common issues seem to be:

- Decision-making delays.
- Organizational structure complexities, too many layers of decision-making and unclear paths for approvals.
- Ambiguities in who has what level of delegated authority.
- Lack of accountability for sustainability performance.
- Failing to address issues and risks head-on as soon as they become apparent.
- Groupthink and optimism bias, or conversely a strong bias towards risk aversion.
- Lack of transparency, weak expectation management, with too much focus on politicking and message management.
- Being inappropriately influenced by internal and external pressures and influences.
- Inadequate organizational visibility, interest, support and constructive challenge.
- Ineffective understanding and management of risk.

How to overcome common issues affecting successful delivery of sustainable innovation projects

The essential elements of effective sustainable innovation project governance are: alignment to wider organizational strategy and corporate governance; risk management; clear decision gates and processes; delegated authorities; effective reporting; independent assurance; and change management. Making sure that your organization has all these elements in place will help to remove or mitigate any project governance issues that might negatively impact your sustainable innovation initiative.

1 Alignment to wider organizational strategy and corporate governance

The strategic management process in an organization involves identifying and executing its long-term objectives. Selecting and successfully executing sustainable innovation projects, whether individually, in programmes or as part of portfolios, is a key activity in achieving these long-term goals. At the strategic level, it's crucial for an organization to set up appropriate governance structures to oversee sustainable innovation project management. Corporate governance encompasses the broader organizational control, within which sustainable innovation project governance is a specific focus area. Sustainable innovation project management, therefore, includes aspects of project governance relevant to the project level, and the management of project details.

2 Risk management

Governance includes identifying and ensuring effective management of or preparation for all types of key risks that could significantly impact the sustainable innovation project:

- Technical risks, such as the feasibility and functionality of the proposed innovation, are usually the most obvious risks. Teams need to question whether their ambitious technological or technical goals can be achieved within the scope and constraints of the project, as explored in Chapter 7. Risks of the sustainable innovation failing to perform as anticipated in a live, less predictable, real-world environment are also important to consider.

- Financial risks include more than just the potential for budget overruns. They extend to the stability and continuity of funding streams and the market, which can be volatile. Market risks, of course, are closely connected to financial ones, particularly with regard to how swiftly the sustainable innovation will be adopted, competitor reactions and volatility and uncertainty of costs, materials and skills availability.
- Regulatory and compliance risks also need to be planned for and managed.
- Social and ethical risks include public perception, reputation and societal impact. Sustainable innovations need to align with social values and strive towards enhancing societal well-being without inadvertently deepening social inequalities.
- Operational risks could also impact the continuity of a sustainable innovation project, with supply chain disruptions and the integrity of internal processes being potential stumbling blocks.
- Strategic risks demand that the sustainable innovation project remains in step with the overarching goals of the organization and features enough flexibility to pivot if those objectives evolve.
- There is also risk associated with human resources, from attracting and retaining top talent to fostering a collaborative and effective team dynamic.
- Technological advancement brings with it the risk of obsolescence, where today's innovation could be tomorrow's relic, overtaken by newer advancements. The challenge extends to ensuring seamless integration of the new technology with existing systems, a task often easier said than done.
- Intellectual property risks, including the protection of the sustainable innovation and the avoidance of infringement on others' patents, are important considerations.

- Project management risks, such as scope creep and stakeholder management, require vigilant oversight to prevent derailment.
- Political instability, extreme weather events and changing trade situations can alter the trajectory of a sustainable project overnight, demanding a proactive and prepared approach to risk management.

Building on the exploration of constraints and risks outlined in Chapter 7, it is essential to transition into more formalized governance protocols for sustainable innovation projects. A critical component of this process is the maintenance of a comprehensive, current and well-understood risk register.

STEP-BY-STEP GUIDE TO IDENTIFYING AND MANAGING RISKS IN SUSTAINABLE INNOVATION PROJECTS

Here is a step-by-step guide to collating, scoring, prioritizing, mitigating, managing and tracking risks in sustainable innovation projects. By following these steps, teams can ensure not only that risks are understood and managed but also that the governance of the sustainable project remains robust and dynamic, capable of adapting to changing circumstances.

- Step 1: Risk identification

  The first step is to establish a comprehensive document that sets out all the risks – i.e. a risk register. This involves brainstorming sessions with project stakeholders, including project managers, team members, external experts and end users, to identify potential risks across all categories such as technical, financial, market and more. Use the different types of risk described in the previous section to prompt ideas. Document all the risks you identify; also record all your work on steps 2–9 below.
- Step 2: Risk analysis

  Analyse each risk to understand its potential impact. Score each one for its severity of impact and likelihood.
- Step 3: Risk prioritization

  With risks scored, they can be prioritized. This is typically done by ordering them from the highest to the lowest based on their score. The

Pareto principle, or the 80/20 rule, can be applied here, suggesting that 20 per cent of the risks will typically account for 80 per cent of the potential impact. These become the focus for detailed planning.

- Step 4: Risk mitigation planning

  For each high-priority risk, develop a mitigation plan. This should outline the steps to either reduce the likelihood of the risk occurring or limit its impact if it does occur. It may involve contingency planning, alternative strategies, resource reallocation or investing in additional research and development.

- Step 5: Risk ownership

  Assign an owner for each risk. The risk owner is responsible for monitoring the risk and implementing the mitigation plan. Ownership ensures accountability and that each risk is actively managed throughout the sustainable innovation project lifecycle.

- Step 6: Risk response implementation

  Implement risk responses as per the mitigation plans. This could involve proactive steps taken immediately or plans that are held in readiness to deploy if certain triggers are met.

- Step 7: Risk monitoring and reporting

  Establish a system for monitoring risks. This should include regular risk review meetings and updates to the risk register. Monitoring ensures that new risks are identified in a timely manner and that mitigation plans are effectively controlling known risks.

- Step 8: Risk communication

  Keep all stakeholders informed about the status of risks and the steps being taken to manage them. Transparency helps maintain trust and ensures that everyone is prepared to act if a risk materializes.

- Step 9: Risk review and adaptation

  On a regular basis, review the risk management process itself. This review should evaluate how risks have changed, whether new risks have emerged and if the mitigation strategies are effective. Adapt the risk management plan accordingly.

3 Decision gates and processes

Decision gates are designated checkpoints within a sustainable innovation project's lifecycle. At these points, formal decisions are made to either grant permission for the project to proceed or renew its existing authorization. These gates serve as structured control points to assess the project's progress and viability. Recording and communicating decisions made at approval gates is important.

4 Delegated authorities

Effective governance in project delivery balances the team's autonomy in risk management with the board's and sponsor's need for control.

Appropriate and clear delegated authorities, i.e. the transfer of decision-making power from a higher authority (like senior management or a board) to individuals or groups at lower levels within an organization is required. This delegation is usually formalized through policies or frameworks that outline the scope and limits of the delegated powers.

Delegating authority for your sustainable innovation projects empowers your teams or individuals closer to the project to make decisions quickly, without needing approval. This agility is essential when rapid responses to new information, changing conditions or unexpected challenges are needed. Also, teams working directly on sustainable innovation projects often have specialized knowledge or expertise. Delegating authority to these teams ensures decisions are informed by this expertise, which should lead to more effective and technically sound outcomes.

Authority goes together with individual and team accountability. It's an important element in gaining clarity on what the sustainable innovation project needs to achieve and for delivering against that brief, as well as managing the risks related to them.

5 Effective reporting

Effective governance reporting is an essential element of sustainable innovation projects. It ensures transparency and accountability and provides a comprehensive view of the project's progress and alignment with strategic objectives. While every organization and sustainable innovation project is different, here is an example of the balanced scorecard approach, integrating relevant key performance indicators (KPIs), for governance reporting.

The balanced scorecard is a strategic planning and management system used for aligning business activities with the vision and strategy of the organization. It enhances internal and external communications and monitors organizational performance against strategic goals (Kaplan & Norton, 1992).[8]

The balanced scorecard for sustainable innovation projects should include a comprehensive set of KPIs in these categories: financial, customer, internal process, learning and growth and, of course, sustainability. Examples of KPIs that each category could include are:

- Financial KPIs such as return on innovation investment (ROII) and budget variance provide a snapshot of the project's economic impact and financial management.
- Customer-centric KPIs, like customer satisfaction index (CSI) and market share growth, reflect the market's response to innovative outputs.
- Internal process indicators such as time to market (TTM) and project milestone compliance show the efficiency and adherence to project timelines.
- Learning and growth metrics such as employee engagement and knowledge management efficiency highlight the project's human and intellectual capital contributions.
- Sustainability KPIs should be selected to reflect and assess the innovation project's goals and success measures, such as:
    - Carbon footprint reduction
    - Waste reduction
    - Energy efficiency improvement
    - Social impact score

It's important to ensure that your KPIs are both leading and lagging. Lagging KPIs are retrospective. They measure the outcomes of actions that have already happened. They're useful for understanding results and outcomes, but they don't provide insight into future performance or give you the means to change the outcome. Some examples of lagging KPIs related to sustainable innovation include:

- **Carbon footprint:** This measures the total greenhouse gas emissions caused by the project, calculated after project activities have taken place. It's a lagging indicator because it can only be measured after the fact.
- **Water usage reduction:** The amount of water saved because of the innovation project, compared to baseline figures, is a lagging KPI that reflects the project's impact on resource conservation.
- **Waste reduction percentage:** Post-implementation, this measures the actual reduction in waste generated by the new process or product, indicating the effectiveness of sustainability initiatives.
- **Actual vs. projected energy savings:** After the project is implemented, this KPI compares the projected energy savings based on the leading KPIs to the actual energy savings achieved.
- **Sustainability-related cost savings:** This is a measure of the financial savings gained from implementing sustainable practices, such as reduced material costs or lower energy bills, realized after the project's completion.

Leading KPIs, on the other hand, are predictive. They indicate future performance and are measured before the outcome is known. These KPIs can influence change because they measure activities that affect future performance.

- **Energy consumption rate:** This measures the amount of energy projected to be used by new processes or technologies during the development phase. It's a predictive indicator of how energy-efficient the innovation will be.
- **Sustainability training hours:** The number of training hours dedicated to educating the project team on sustainability practices can indicate how well prepared the team is to implement sustainable methods.
- **Percentage of recycled materials used:** For projects involving manufacturing or construction, this KPI tracks the incorporation

of recycled or sustainable materials in the early stages of product development.

- **Innovation investment on sustainability:** The proportion of the project budget allocated to sustainable practices or technologies can be a leading indicator of the project's commitment to sustainability.
- **Supplier sustainability assessment scores:** If the project involves external vendors, this KPI assesses their sustainability practices, indicating the potential sustainability of the supply chain.

A balanced mix of leading and lagging KPIs provides a more comprehensive view of performance. Leading KPIs allow adjustments before it's too late, while lagging KPIs are essential for assessing the results of project activities. By monitoring both, you can manage and improve performance proactively rather than simply reacting to past results.

Governance reporting will need to be dynamic, with KPIs and targets adaptable to the evolving landscape of the project. The frequency of reporting should be determined with an eye towards providing timely information without overwhelming the project team or stakeholders. Each report should serve as a springboard for continuous improvement, helping the team to deliver the sustainable innovation successfully.

A well-structured balanced scorecard enriched with a dedicated sustainability perspective provides a robust framework for governance reporting in sustainable innovation projects. It ensures that all areas of performance are scrutinized, from financial viability to social and environmental impact, forming a comprehensive picture of the project's progress. As these projects forge the path towards a more sustainable future, the insights gleaned from effective governance reporting will be invaluable in guiding strategic decisions and ensuring responsible stewardship of both organizational and natural resources.

6 Independent assurance

Assurance is a methodical approach designed to give senior leaders and key stakeholders the confidence that a sustainable innovation project is well managed, on course for delivery and in alignment with organizational policy or strategy.

For effective governance, it is essential for sustainable innovation projects to incorporate a clear, consistent, independent and comprehensive assurance and approval plan within their governance frameworks. This plan should be established at the project's outset, with regular evaluations and adjustments carried out until project completion.[9]

Assurance activities are checks and balances designed to ensure that sustainable innovation projects meet their goals, adhere to standards and align with ethical and environmental guidelines. They play a crucial role in project governance, providing confidence to stakeholders that objectives will be achieved responsibly.

The key functions of assurance activities include quality control, compliance with regulations and ethical standards, risk management and performance monitoring.

Assurance activities should be structured across three distinct levels:[10]

- **First level:** Executed by or on behalf of the operational team directly accountable for managing risks. This level ensures that the sustainable innovation project adheres to the requisite standards, including those related to environmental and social benchmarks. Typically, this is the responsibility of the sustainable innovation project delivery team.
- **Second level:** Conducted by individuals or groups within the project or organization who are not part of the first-level activities. Their role is to verify that the first level of defence is adequately structured, active and functioning as planned. Often, this assurance is provided by an internal management function that offers an unbiased assessment, as they are not directly responsible for the risks being assured.
- **Third level:** Performed independently to offer senior management an impartial evaluation of the efficacy of governance, risk management and internal control systems, which encompasses both the

first and second lines of defence. This level is typically carried out by entirely independent teams external to the project and equipped with the necessary expertise to conduct thorough project analyses. Their independent review should also rigorously scrutinize environmental and social safeguards.

Assurance reviews should be strategically timed prior to significant project milestones, such as approval gates, to provide decision-makers with a current evaluation and future projection of the project's trajectory. Planning for these reviews should balance thoroughness with the need to minimize disruption to the delivery team, management and reviewers.

The need for challenge and support goes beyond usual assurance methods and processes, however. In sustainable innovation, projects often push the boundaries of conventional practice and navigate uncharted territories, meaning that independent challenge is crucial.

Independent challenge involves experts, colleagues or consultants who are not involved in the day-to-day operations of the innovation. Their role is to review project plans and outcomes with a critical eye and question underlying assumptions, methodologies and the viability of proposed solutions. It's also to offer alternative perspectives based on experience in other sectors or industries and validate the sustainable innovation project's alignment with broader sustainability goals and principles. Ultimately their role is to ensure the project's adherence to external standards and stakeholder expectations.

Independent challenge brings a level of scrutiny and perspective that counters some of the natural biases and assumptions that feature in even the most dedicated and experienced project teams. Other benefits are that knowing that an external challenge will occur encourages project teams to elevate their standards of proof and argumentation, fostering a culture of rigour and thoroughness. Challengers can help identify risks earlier in the project lifecycle, which can be mitigated more cost-effectively than if discovered later. Plus, projects that have undergone rigorous independent challenge are more credible to stakeholders, including investors, regulators and the public. This credibility is vital for securing funding and support.

Finally, independent reviewers often bring insights from other industries and disciplines, which can be invaluable in breaking through innovative barriers and facilitating cross-industry innovation.

While challenge is necessary, support is equally vital. External supporters can offer a sounding board for developing and helping to test new ideas and proposed approaches. They can bring specialized knowledge in sustainable practices, technology or market trends that the internal team may lack. Expert supporters should include seasoned professionals who have navigated similar innovation challenges. Experts with connections to potential partners, customers or resources that can aid in project development are useful. External experts can also be helpful in validating the project's sustainability claims, which can be essential for marketing and scaling the innovation.

7 Authorizing and monitoring sustainable innovation change management processes

Project change management disciplines are essential for sustainable innovation governance. It's important to create and maintain a carefully controlled environment for change, ensuring that each modification to the project's plan is thoroughly vetted and approved before being put into action.

Changes are bound to be necessary at various stages of the sustainable innovation process due to factors such as new learning, emerging technologies, shifting market conditions or evolving regulatory landscapes. Projects need to be flexible enough to adapt to change, while continuing towards the original vision and objectives. An effective project change process helps to prevent scope creep, irrelevance, time and budget issues, and more. It also ensures that change decisions are made by the right people, working at the right level, and that the sustainable innovation doesn't diverge too far from what was promised and is expected by key stakeholders.

The change control system implemented needs to be robust and overseen by a designated leadership team, who should review every proposed significant alteration. Clear and specific procedures for submitting change requests are critical to the effectiveness of this process. They should require an adequate description of the change, its justification and its anticipated impact on the project. Every material change request should undergo a rigorous evaluation to weigh its necessity against potential benefits and costs, while also assessing the risks and impacts on the project's deliverability. The leadership team that makes up the governance body should either approve or veto these changes, based on its assessments. Of course, the change process needs to align with the organization's delegated authorities, so that decisions can be taken at the right level, without unnecessarily slowing progress.

Implementing changes will also require updates to project plans, budgets and schedules. These impacts will need to be communicated transparently to stakeholders to maintain trust and alignment. Importantly, each change, along with its rationale and outcome, should also be documented. This ensures clarity and accountability as well as building a knowledge base for future innovations. Regular reviews of the change management process are essential to refine its effectiveness and to ensure that it continues to serve the sustainable innovation's best interests.

More advanced governance considerations for larger-scale, highly complex sustainable innovation programmes and projects

The fundamentals of effective project governance described in this chapter are necessary for all sustainable innovation activities. If you're working at scale, with high degrees of complexity that also might involve incorporating innovative technologies, materials and/or methods, they are especially important. Your governance strategy for all sustainable innovation projects needs an innovative and contemporary approach, especially with respect to risk allocation and consideration of how to position your organization as a client or supplier of choice.

Many sustainable innovation governance models are still based on traditional ways of working, which are adequate for projects that are reliant on repeated, known and well-understood processes, technologies and materials that feature limited levels of innovation. These do not work well for more ground-breaking projects that involve the client and supply chain partners innovating together. There will likely be some internal and upwards education of key decision-makers and stakeholders, and a new mindset and framework when it comes to risk are now needed.

Existing models often aim to push risk from the client onto the supplier, where possible. This can be priced and agreed effectively when performance and outcomes are reasonably predictable because innovation is not a significant project feature. But when the risks are much less certain and sit further outside the supplier's control, a new way of collaborating, risk-sharing and contracting will usually be needed to allow the project to progress on terms that are not prohibitively expensive for the client. Suppliers, as well as clients, have their own governance and risk tolerance to consider, too. Clients also have a moral, as well as practical, obligation to take reasonable steps to help their supply chains to survive and thrive.

This collaborative, risk-sharing, joint-problem-solving and innovative approach requires much more advanced leadership and commercial and project management skills from teams than the traditional, transactional approach to managing supplier relationships. Client organizations need to build greater capability to fully understand the new risks of their sustainable innovation projects, and the controllability, impact, cost premia and effect on performance of these risks, to learn to work with suppliers in new ways to achieve mutually beneficial outcomes.

CASE STUDY

A transformation model for the water sector in England and Wales

Since its privatization in 1989, the water sector in England and Wales has achieved some improvements in pollution reduction, resilience, water quality, service levels and efficiency. The supply chain has been integral in these advancements. However, from its next business plan cycle, which starts in 2025, this highly regulated sector will experience intensified challenges such as affordability, customer service expectations and environmental and climate resilience needs.

By 2023, the industry's supply chain showed signs of increased fragility, highlighted by the collapse of a key tier-one supplier. Issues such as capacity constraints, inflationary pressures and more lucrative opportunities in other sectors also highlighted some of the other potential risks for water companies seeking to make the future changes they have committed to, overseen by industry regulator Ofwat.

In 2023 and 2024, the sector began to procure for its 2025 projects, against a backdrop of rising costs due to inflation, Brexit, the aftermath of the Covid-19 pandemic, geopolitical events and increased demands from other sectors. There was also a noticeable skills shortage and market capacity reduction. The water sector's low profit margins further exacerbated the issue, making working for the sector less attractive to some suppliers.

As a result, water companies reassessed their risk allocation strategies to become 'clients of choice' to attract suppliers amid these challenges. This involved developing a new governance mindset, creating different contract models, enhancing client-side capabilities, finding ways to manage new risks and updating KPIs and incentive models. Water companies also began to build internal capacity and capability to support these models and help to support a sustainable supply chain ecosystem.[11,12]

For example, Northumbrian Water sought interested parties for its £8 billion capital delivery frameworks in construction and engineering from 2025 to 2030. These frameworks have the potential for a 12-year extension through additional negotiations with the chosen suppliers. The company has highlighted a need for a significant shift in collaboration, risk management and capital investment coordination for its 2025 programme. To realize this, Northumbrian Water crafted a comprehensive strategy and initiated an extensive transformation programme to found its innovative Living Water Ecosystem.

The company's delivery strategy included two main pathways, tailored to the project's size, complexity and need for technical expertise. These were defined as short cycle and long cycle work.

Short cycle work encompassed swift and flexible capital investment in water and wastewater assets, including infrastructure services for development clients. Long cycle work entailed more intricate projects, predominantly managed by Northumbrian Water's Living Water Enterprise (LWE). This enterprise aims to operate as a cohesive, outcome-focused unit with a strong emphasis on collective success over individual financial gain.

The LWE is committed to:

- Unifying all involved parties under shared objectives and outcomes.

- Prioritizing performance incentives to ensure success is a common goal.
- Integrating totex (total expenditure) principles from the beginning.
- Focusing on delivery optimization at a programmatic level to employ a 'best athlete' approach, avoiding strict supplier divisions.

Northumbrian Water also created opportunities for firms new to the water industry, understanding the importance of fostering new industry capacity and aiding new market participants.

Key roles

The ecosystem of project governance in sustainable innovation involves three key roles: the project sponsor, project manager and project management office (PMO). Each has a distinct yet important role to play in governance, decision-making and project progress.

Project sponsor

A project sponsor champions the project, providing not just the financial backing but also the organizational support needed for the sustainable innovation to succeed. They are responsible for ensuring that the targeted value from the project is achieved, articulating the vision and sustainability and other objectives of the project. They also must ensure that it's not an isolated endeavour but a strategic fit for the company's broader aims. With their influence, the project receives the advocacy and decision-making influence necessary to navigate strategic challenges and senior stakeholders. It's their role to secure the essential resources and to act as the project's advocate across the organization.

Project manager

The project manager is the operational lead for the sustainable innovation, turning the sponsor's vision into an actionable plan. This role involves managing the team and resources, thus keeping the project on track through adept planning and execution. The project manager becomes the linchpin of communication, linking the sponsor, team and stakeholders together and ensuring everyone is updated and the project's sustainability goals are clear and pursued. Monitoring the project's progress and steering it towards timely and on-budget completion with the desired sustainability and other outcomes are at the heart of this role.

Project management office (PMO)

Lastly, the PMO underpins the project with governance, ensuring consistency and alignment with the organization's project management standards. This office oversees various projects, ensuring they adhere to established sustainability frameworks and organizational standards. The PMO supports project teams by providing them with the necessary tools, training and best practices to manage their projects effectively. Moreover, it has the crucial task of reporting on project performance to senior management, ensuring transparency and accountability. Resource management is another critical function of the PMO, as part of its role is to allocate resources across the organization's portfolio of projects.

Not every organization has a PMO, since its presence often depends on the size, complexity and number of projects an organization undertakes. Typically, larger organizations with a significant portfolio of projects across various departments or teams will establish a PMO to maintain oversight, ensure standardization and optimize project delivery.

Organizations that often find a PMO particularly beneficial include multinational corporations, government agencies and non-profits, especially those operating in industries where sustainable practices are crucial, like energy, construction, manufacturing and technology. This is because they are often running multiple, complex sustainable innovation projects simultaneously, on a continuous basis. For such entities, a PMO can guide the integration of good governance and decision-making into project management practices, ensuring that these values are embedded in every project phase, from initiation to closure.

A PMO can lead in driving sustainable innovation by fostering a culture that prioritizes sustainability within the project management discipline. This also ensures that project managers and their teams have the tools and knowledge necessary to incorporate sustainability into their projects. Moreover, a PMO can track and report on the sustainability outcomes of projects, making sure that they align with organizational goals and, where relevant, regulatory requirements or industry standards.

In essence, the PMO serves as a central hub for best practices, learning and continuous improvement in sustainable project management, which is increasingly vital for organizations seeking to innovate responsibly and effectively.

Other roles

In addition to the project sponsor, project manager and the PMO, there are several other roles that are integral to the governance and decision-making processes in major sustainable innovation projects. Depending on the project, these could include:

STEERING COMMITTEE

Often comprising senior stakeholders, the steering committee provides strategic direction and oversight. They make key decisions that affect the project's trajectory, ensuring that it remains aligned with the organization's sustainability objectives.

BUSINESS ANALYSTS

Business analysts can play a critical role in understanding and translating business needs into project requirements. Business analysts ensure that the innovation is designed to deliver value and meet the sustainability and other goals set out by the organization.

TECHNICAL EXPERTS/SUSTAINABILITY SPECIALISTS

Individuals with expertise in sustainable practices provide critical insights into the technical aspects of the innovation project. They ensure that sustainability is not just a checkbox but integrated into the project's fabric through innovative solutions and practices.

RISK MANAGER

The risk manager role involves identifying, analysing and mitigating risks, particularly those that could impact the sustainability outcomes of the project. They help in devising strategies to deal with potential issues before they affect the project.

CHANGE MANAGER

The change manager is responsible for managing the organizational change process, ensuring that the project's sustainability initiatives are adopted and embraced across the organization. Not to be confused with the project change management process discussed earlier, this role is specifically concerned with making sure that the innovation, once complete, becomes adopted successfully as business-as-usual.

PROJECT TEAM MEMBERS

The individuals who execute the project tasks, informed by the project's sustainability objectives, play a direct role in decision-making at the operational level.

STAKEHOLDERS

Stakeholders can include clients, end users, community representatives and suppliers, among others. Their input is valuable for shaping project outcomes that are sustainable and meet the needs of all parties involved.

Conclusion

Good governance and smart decision-making are critical for the success of sustainable innovation projects. Not only are they essential for the internal workings of an organization but they also meet the expectations of regulators, customers and investors. Governance within an organization should mirror its broader corporate strategy while offering a balanced framework that emphasizes accountability and thoroughness without stifling flexibility or innovation with unnecessary red tape.

The governance framework for sustainable innovation must be well defined, ensuring that everyone involved understands the decision-making process and authority levels and adheres to them. Risk management is also a vital element, necessitating a tailored risk register and action plan for each project, which should be regularly updated to address any changes or new insights.

Reporting is another key aspect, with a blend of lagging and leading KPIs providing insights into the project's health and trajectory, helping guide decisions at critical junctures. To maintain project integrity, there should be regular assurance checks at all delivery levels, bolstered by independent expert review.

When dealing with complex projects that require deep collaboration with supply chain partners, there must be a shared commitment to innovation and equitable risk sharing. The effectiveness of these partnerships is mirrored by the competence of the project management team, from the project sponsor to the team members, ensuring a robust and skilled approach to project execution.

ACTION CHECKLIST

- Review the structure of the governance structure and processes for your sustainable innovation programmes and projects. Carry out a strengths and gaps analysis against the criteria set out in this chapter:
  - Alignment with corporate governance.
  - Appropriate decision gates, delegated authorities, reporting and KPIs, change process, assurance, independent challenge and support.
  - Effective, regularly updated risk register and action plan.
  - Appropriate approach to risk allocation, contracting and collaborative working with supply chain partners.
  - Good project team, with all key roles covered as needed, with the right people doing the right things at the right time to the right standard.

- Consult with key stakeholders and decision-makers to inform your review.
- Based on your review, create and implement an action plan for improvement.

Notes

1. Chartered Governance Institute UK & Ireland (n.d.) What is corporate governance?, www.cgi.org.uk/about-us/policy/what-is-corporate-governance (archived at https://perma.cc/9CLD-VRJH)
2. CIPD (2023) Corporate governance: An introduction, www.cipd.org/uk/knowledge/factsheets/corporate-governance-factsheet/ (archived at https://perma.cc/TA6Q-S35G)
3. Camara, A (n.d.) The importance of ESG in corporate governance, Corporate Finance Institute, https://corporatefinanceinstitute.com/resources/esg/importance-of-esg-in-corporate-governance/ (archived at https://perma.cc/VLR5-R56G)
4. PwC (2021) PwC's 2021 Global investor survey, www.pwc.com/gx/en/services/audit-assurance/corporate-reporting/2021-esg-investor-survey.html (archived at https://perma.cc/SY9G-AZAU)
5. Federal Trade Commission (n.d.) Environmentally friendly products: FTC's Green Guides, www.ftc.gov/news-events/topics/truth-advertising/green-guides (archived at https://perma.cc/RQK7-RRHB)
6. Maersk (2023) Sustainability, www.maersk.com/sustainability (archived at https://perma.cc/LLK3-JGB8)
7. Infrastructure and Projects Authority (2023) Project Development Routemap for Infrastructure Projects: International Module, Governance, GOV.UK, www.gov.uk/government/publications/project-development-routemap (archived at https://perma.cc/TR7K-LCYV)
8. Kaplan, R S and Norton, D P (1992) The balanced scorecard – measures that drive performance, *Harvard Business Review*, https://hbr.org/1992/01/the-balanced-scorecard-measures-that-drive-performance-2 (archived at https://perma.cc/63CM-86NG)
9. HM Treasury and Cabinet Office (2011) Major Project approval and assurance guidance, https://assets.publishing.service.gov.uk/government/uploads/system/uploads/attachment_data/file/179763/major_projects_approvals_assurance_guidance.PDF.pdf (archived at https://perma.cc/34UD-46NC)

10 HM Government (2021) GovS 002: Project delivery portfolio, programme and project management, https://assets.publishing.service.gov.uk/government/uploads/system/uploads/attachment_data/file/1002673/1195-APS-CCS0521656700-001-Project-Delivery-standard_Web.pdf (archived at https://perma.cc/V9L3-X3SA)

11 ACE (2022) Water Industry Forum report informs thinking for AMP8 delivery models, www.acenet.co.uk/news/industry/water-industry-forum-report-informs-thinking-for-amp8-delivery-models/ (archived at https://perma.cc/293R-B93U)

12 Environment Analyst (n.d.) Record water investment predicted for AMP8, https://www.rand.org/pubs/reports/R4246.html (archived at https://perma.cc/CRH6-9PPA)

09

Building a sustainable innovation strategy: Reimagining the innovation pipeline

CHAPTER OVERVIEW

Building a sustainable innovation strategy and reimagining the innovation pipeline are at the centre of Chapter 9. Here, you'll be guided through a step-by-step process on how to develop a sustainable innovation strategy, with a focus on reimagining the innovation pipeline and a process to accommodate the unique demands of sustainable innovation.

The steps for creating a sustainable innovation strategy start with taking the horizon-scanning and stakeholder engagement insights that you have generated from Chapter 3, complementing these with a competitor review and a performance analysis of your organization's infrastructure, projects, products and services in terms of sustainability, customer satisfaction and growth trajectory. This data will inform your sustainable innovation diagnostic to identify your organization's current position, needs and sustainable innovation opportunities. The individual and collective level of sustainable innovation ambition for these opportunities can be mapped to visualize the spread of opportunities and the overall risk profile of the strategy.

Building a sustainable innovation pipeline comes after creating the strategy. The strategy provides a range of opportunities and problems to solve. The sustainable innovation pipeline helps to organize individual items into projects and programmes, and crucially prioritizes them based on clear criteria, to ensure that attention, resources and money are invested in the most appropriate activities for your organization.

These prioritized projects will then need to be progressed using the sustainable innovation process, or funnel, which optimizes as efficiently as possible the probability of making the most promising innovations a reality as efficiently as possible. This incorporates a well-informed stage-gate decision-making process. Barriers to sustainable innovation progress need to be removed or mitigated wherever possible, and projects measured and tracked as they fall out of the funnel or move through it. The sustainable innovation strategy and pipeline need to be constantly updated with any significant changes, and the organization needs to have the capability for an agile response to those changes.

From opportunity to reality: A holistic approach to making sustainable innovation happen

Horizon-scanning and stakeholder engagement ideally will drive out numerous sustainable innovation challenges and opportunities for the organization, but the likelihood is that there are many others to be discovered and that need to be considered before strategic decisions are made.

Looking outwards, at the world the organization is and will be operating in, is a healthy place to start for leaders of sustainable innovation. It provides an important context for the effective analysis of competitor and internal landscapes, supporting the creation of a future-ready sustainable innovation strategy, and from there moving the most needed and promising activities from problems and ideas into action.

Sustainable innovation strategy

A sustainable innovation strategy is a plan that guides leaders in developing and implementing innovations that are environmentally responsible and economically viable, while aligning with the organization's mission, vision and values. It involves making strategic decisions based on clear criteria, prioritizing goals that ensure long-term sustainability and efficiently using resources to minimize environmental impact. The strategy should be for a targeted future period, depending on the sector, pace of change, investment cycles and projected lifespan of the organization's products, infrastructure and technologies. The sustainable innovation strategy therefore needs to be constantly updated with any significant changes, such as technological developments, infrastructure changes or new requirements, and the organization needs to have the capability for an agile response to those changes.

A sustainable innovation strategy is a critical tool for leaders of large-scale, complex sustainable innovation initiatives. Developing, communicating and implementing a clear strategy has numerous benefits. It means that decision-making can be based on clear priorities, parameters and criteria set out in the strategy. As economist and strategist Michael Porter wrote, 'The essence of strategy is choosing what not to do.'[1] There will always be more opportunities and challenges than organizations have the time, energy and resources to address, so the focus that having a strategy provides is invaluable. The sustainable innovation strategy is also a useful communication and engagement tool for stakeholders, customers, employees and suppliers. Internally, it facilitates cross-functional collaboration and shared objectives. Having and working to an effective sustainable innovation strategy is a way of keeping the organization relevant, current and future-ready, which reduces performance and financial risk.

Your sustainable innovation strategy is a detailed plan that outlines:

- The specific sustainability goals that your organization aims to achieve.
- The value propositions and business models of your sustainable innovations.
- A course of action for enhancing legacy assets.
- The operational shifts required to implement these innovations effectively.

- The data, infrastructure and technology upgrades necessary to support sustainable practices.
- A talent and resource allocation plan that aligns with your strategic sustainable innovation goals.
- Clear financial projections that factor in the investment and potential returns of your sustainable innovations.
- A timeline for implementation, with milestones for review and adjustment.

Your strategy should be a clear blueprint that guides decision-making and action across the organization, with an emphasis on measurable outcomes and accountability. It should include distinct, actionable steps for your organization to follow to achieve its sustainability and innovation objectives.

How to build a sustainable innovation strategy

Your sustainable innovation strategy is the outcome from your comprehensive collation of insights from your horizon-scanning and stakeholder engagement (Chapter 3), value proposition and business model (Chapter 4) and the findings from your review of existing legacy assets (Chapter 5), all informed by data-based evidence (Chapter 6). Depending on the nature of your organization, its sector, mission and vision, and sustainable innovation project, you may also wish to complement these inputs with a competitor analysis. Competitor analysis evaluates the strengths and weaknesses of your competitors' current performance and strategies to help identify opportunities and threats for your business to enhance your own strategy. It's about assessing how much better or worse your organization is at helping your customers to achieve their objectives.[2] It is also not about copying – it's about outperforming, finding gaps and opportunities for sustainable innovation and growth.

Here's how to do a competitor analysis for your sustainable innovation projects in six steps:

1 Identify your competitors.

2 Review how your products and services compare with those of your competitors.
3 Work out what additional intelligence you need on your competitors and why.
4 Get the competitor intelligence you need.
5 Combine the competitor intelligence with customer needs and trends.
6 Identify gaps and potential opportunities for innovation and competitive advantage in your own organization.

You will also need to do an internal review to identify how well placed your organization is to rise to the challenges and opportunities ahead. Any development actions you'll need to take for sustainable innovation readiness should be included in your sustainable innovation strategy. For each of the following areas, map out your current and desired position:

- Using available data, analyse the performance of each of your current products and services in terms of their sustainability, sales volume, value and profitability, customer satisfaction and growth trajectory. Identify those high potential or critical products and services that you want to invest in heavily, those that you need to maintain and those that you will deprioritize or even remove completely.
- Assess how fit your key business operations and processes are for delivering your current and future sustainable innovation needs. Do the same for your infrastructure and technology assets. Base these assessments on sustainability, user satisfaction, efficiency and effectiveness, including the potential return on investment of any improvements you deem necessary.
- Review the capacity, capability, effectiveness and culture of your team, supplier base and ecosystem. Chapter 10 delves into more detail on these themes.

Your organization's strategic choices, meaning the decisions it makes about its long-term goals and the direction it plans to take to achieve these goals, involve selecting the most appropriate path from a range of potential strategies after a thorough analysis of internal and external factors. Clearly summarize the key, potential strategic choices available to your organization in terms of external opportunity and internal improvement. Map these potential choices onto the levels of innovation ambition below, which categorize innovations based on how core or transformational they are compared with your present activities.[3]

Levels of innovation ambition:

- Core innovations are incremental. They focus on optimizing current products, services, operations, processes, infrastructure, technology and other assets for current customers and users. These are usually innovations of lower risk, investment requirements and complexity.
- Adjacent innovations are newer to the organization than core innovations. They often share some of the characteristics of core and transformational innovations, but their focus is on tapping into the organization's strengths and applying them in new arenas. Risk, investment requirements and complexity tend to be moderate.
- Transformational innovations are potentially game-changing for the organization and target significant breakthroughs. They involve innovating brand-new solutions and creating a market for them. Transformational innovations generally involve high risk, investment requirements and complexity.

Mapping your potential strategic sustainable innovation choices to these three categories shows the overall risk profile of all of them combined. A 'golden ratio' of 70 per cent core,[4] 20 per cent adjacent[5] and 10 per cent transformational innovation[6] is often cited as a yardstick for a reasonably balanced distribution of risk, reward and cash flow maintenance, as well as for supporting the organization's capacity to deliver innovation on top of running day-to-day business operations. This means a strong base of reliable, low-risk innovations that ensure steady revenue, while also investing in more speculative ventures that could lead to significant future growth and market differentiation.

For an infrastructure example, the levels of innovation ambition might be as follows:

- 70 per cent core innovations – the bulk of the portfolio focused on enhancing and optimizing existing infrastructure with relatively small changes. This might include upgrading to energy-efficient lighting, improving insulation in buildings to reduce heating and cooling needs or retrofitting older structures with smart sensors to monitor health and usage, all aimed at reducing environmental impact and saving costs.
- 20 per cent adjacent innovations – extending the organization's current capabilities into new but related areas. These innovations might involve developing new materials that are more sustainable and have a lower carbon footprint, or branching into the creation of infrastructure for new markets or uses.
- 10 per cent transformational innovations – breakthrough, market-creating or leading-edge innovations. An example could be investing in the research and development of carbon-negative building materials.

However, there is genuinely no one size that fits all. The ratio between the levels of innovation will need to vary according to the organization's sector, lifecycle stage, financial position and capabilities.

Balancing sustainability, affordability, value, time and risk (Chapter 7) inform these strategic choices, which are overseen by effective sustainable innovation decision-making and governance processes (Chapter 8).

Sustainable innovation pipeline

The sustainable innovation pipeline is the bridge between strategy and implementation, setting out what innovations will be delivered when, based on the organization's priorities. The pipeline should also capture additional innovation opportunities as they occur, so that they can be ready for prioritization. As innovations are completed and time passes, ideally there will be a continuous feed of opportunities entering the pipeline.

It is unlikely, owing to limitations of time, resources and funding, that all the strategic sustainable innovation choices made by the organization can be implemented simultaneously. Plus, some will be more desirable, feasible, viable[7] and sustainable than others. Leaders of sustainable innovation therefore need to prioritize which innovations will be worked on, and when, balancing the achievement of quick wins with longer-term yet important projects. As discussed in Chapter 8, having a clear methodology and criteria to determine the contents of the pipeline provides an objective and collaborative framework for decision-making. But, as with all sustainable innovation decision-making, the initial investment of money, time and effort should be sufficient to guide the innovation process effectively, but not so much that it becomes wasteful, a barrier to overall progress or unsustainable.

Defining the strategic choices

Each strategic choice needs to be clearly defined as follows to ensure that there is a shared understanding of the opportunity or challenge:

1. Description of the sustainable innovation.
2. What is the value proposition (Chapter 4) for this innovation for the organization, customers, stakeholders and users?
3. Assessment of the sustainable innovation's desirability, feasibility, viability and sustainability:
   - Desirability means that there is early evidence of sufficient demand or requirement for the sustainable innovation from customers and stakeholders (external projects) or employees (internal projects).
   - Feasibility is the practicality of turning an idea into a working solution. It considers the technical capabilities, the availability of technology, resources and the time required. If a project is feasible, it means it can be created and delivered with the expected means and existing constraints.

- Viability is about the economic logic of a sustainable innovation. A viable product or service is one that is not only demanded by the market or required internally but can also be produced and delivered at a cost that is profitable or sustainable for the organization. Viability assesses whether the business model supports long-term growth and profitability.
- Sustainability, as we have seen, evaluates the long-term effects of the strategic choice on the environment and society. This includes looking at resource use, energy consumption, waste generation and the potential for recycling or reuse.

4. Categorization of the sustainable innovation as a core, adjacent or transformational innovation.
5. Initial estimates of:
    - How long the sustainable innovation will take to develop and implement.
    - How much investment will be required, and when.
    - The people and other resources required.
    - Timescales for potential benefits to begin to be realized.
    - Any foreseeable, material risks and constraints (Chapter 7).
6. Confidence level, expressed as a percentage, of the sustainable innovation project succeeding, based on the early information available.

Where there are numerous strategic choices, these should all be captured in a format such as the one shown in Figure 9.1. Items 2–5 above could be scored to enable sorting and prioritization of all the strategic choices.

Once you have assembled all the data for points 1–6 above for each of the potential strategic choices, you can then decide which ones should be included in your sustainable innovation pipeline. The data and scoring will provide some objective input for discussions, with decisions made collaboratively between several key decision-makers with the optimal mix of technical and commercial expertise. Ideally, the individuals or teams who originally worked on the strategic choices will be given the opportunity to 'pitch'; in other words,

FIGURE 9.1 Sustainable innovation pipeline format

	Feasibility	Desirability	Viability	Sustainability	Prioritization score
Sustainable innovation #1	5	5	5	5	20
Sustainable innovation #2	3	4	4	5	16
Sustainable innovation #3	2	4	3	3	12
Sustainable innovation #4	5	2	2	2	11

	Categorization	Investment needed	Resources required	Time to benefit	Confidence level
Sustainable innovation #1	Core	$	Low	Low	95%
Sustainable innovation #2	Core	$$	Low	Low	80%
Sustainable innovation #3	Adjacent	$$$	Medium	Medium	60%
Sustainable innovation #4	Transformational	$$$$	High	High	50%

explain the rationale and answer any questions that decision-makers may have. Collaborative decision-making can reduce risk and lead to greater objectivity by bringing diverse perspectives and expertise to the table. Additionally, when a group collaborates on a decision, there's a check and balance system where biases can be challenged and assumptions can be tested. This can result in decisions that are more balanced and grounded in a broader range of data and experiences, which generally leads to better outcomes. It is likely that, following discussions, some of the content and scoring in the data table may need to change to reflect any insights or observations that emerge.

Should the organization progress with the strategic choice?

For each strategic choice, there are three fundamental decisions to be made when formulating an organization's sustainable innovation pipeline. They are:

1 Should the organization progress with the strategic choice? If the answer is yes, it will be added to the pipeline. A 'maybe' innovation

will need more work or go on the backburner. A 'no' will be recorded, along with the reasons why, and shelved.

2 How does each strategic choice going into the pipeline relate to the others? For example, are there some projects that have a dependency on others; can some be grouped or batched together for efficiency and effectiveness; and should some groups of projects be structured into programmes?
3 What is the optimal way to prioritize implementation of all the projects in the pipeline to derive maximum value for the organization and its stakeholders, balancing quick wins with long-term gains, and optimizing the use of available resources and cash flow?

When a strategic choice is confirmed as entering the pipeline, it becomes a sustainable innovation project to be implemented.

How does each strategic choice going into the pipeline relate to the others?

Understanding the interconnectedness of various projects is crucial. When you have a multitude of projects, seeing how they relate to each other can significantly streamline your implementation. For example, there may be opportunities for smarter procurement, as you can identify overlapping needs and consolidate purchases. It also aids in planning, helping you to schedule projects in a way that either minimizes disruption in each area or concentrates the disruption to deal with it all at once. Some projects may have a dependency on the completion of another project before they can begin. Additionally, by grouping related projects, you can form programmes. Programmes are collections of related projects managed in a coordinated way to obtain benefits and synergies that couldn't be achieved by managing each project in isolation. This consolidation into programmes makes management easier and often leads to more coherent and impactful outcomes.

A project is a singular endeavour with a specific objective, timeline and resources. It's like a single piece in a larger puzzle. A programme, on the other hand, is a collection of related projects. It's the bigger picture that these pieces create when put together. While projects are focused on specific outputs, programmes are oriented towards broader, long-term goals and often involve managing interdependencies and changes across projects. This larger scope of a programme allows for more strategic management and can lead to more significant, sustainable innovation impacts.

There are innovation project management platforms that allow for tagging or categorizing projects. These platforms can help sort projects by various parameters such as location, resources needed or project type, and help with identifying similar projects that could be grouped into programmes. Constructing your sustainable innovation strategy and pipeline from scratch may seem like a lot of work. While it will take some time and effort to set up initially, provided that the organization keeps it updated, it will reap the benefits and maintaining the strategy and pipeline should be straightforward and efficient. If your organization has an existing strategy and pipeline, it may be worth reviewing it to identify areas for improvement.

CASE STUDY

Northumbrian Water's integration of Copperleaf's decision analytics

Northumbrian Water started integrating Copperleaf's decision analytics solution into its asset management planning processes in 2021 to enhance efficiency and sustainability.[8] The company's primary objective was to refine its approach to managing a substantial portfolio of water and wastewater asset projects. The company sought to develop a method that would enable it to evaluate investments comprehensively, ensuring environmental and customer-centric goals were at the forefront of their planning and execution.

The Copperleaf tool assists Northumbrian Water in evaluating their project investments against a unified economic scale, providing a means to quantify and compare the value of disparate investments across the organization. This was a strategic move to bring consistency and clarity to investment decisions. The decision analytics offer a transparent methodology for assessing expenditures across service divisions and facilitates informed decisions for capital, maintenance and operational interventions to better meet the needs of customers and environmental stewardship. The innovation pipeline supports value-based decision-making, enables rapid analysis of multiple scenarios and optimizes investment planning for the water industry's regulatory and delivery periods.

As the impacts of climate change on water supply and infrastructure continue to grow, tools such as Copperleaf are helping water and wastewater companies to optimize assets and build resilience for the long term, while prioritizing the journey to net zero. They also are helping some of these organizations to decide where and when to invest to maximize capital efficiency, meet performance targets, manage risk and achieve their financial and environmental, social and governance (ESG) goals.[9]

What is the optimal way to prioritize implementation of all the projects in the pipeline?

The next fundamental decision is how to prioritize implementation of the projects in the pipeline to derive maximum value for the organization and its stakeholders, balancing quick wins with long-term gains, and optimizing the use of available resources and cash flow; that is, to determine the plan of what to do when, why and by whom. To establish that plan, use the information collated in the data table (i.e. Figure 9.1):

- Examine each project's potential for quick wins versus long-term gains. Quick wins provide immediate benefits, sometimes also helping to set the foundations for more extensive, long-term projects.
- Evaluate the resources required for each project, including personnel, technology and financial investment. Consider the current availability of these resources and how they might be shared or stretched across projects.
- Consider the cash flow implications of each project. Prioritize projects that balance upfront costs with projected returns, to protect financial stability.
- Decide who will be responsible for each project. Choose individuals or teams based not only on their skills but also on their capacity and interest in each project.

- Develop a timeline for all projects that sequences them logically, considering dependencies between projects and other strategic considerations.

CASE STUDY

Tyne 2050: from strategy to action

As we saw in Chapter 6, the Port of Tyne's Tyne 2050 strategy is an intentionally ambitious plan to transform the port and the surrounding region. It launched in 2019 and aligns with governmental economic strategies to promote growth while committing to environmental sustainability. The strategy includes becoming carbon-neutral by 2030, embracing clean energy and advancing in technology and innovation, including establishing the UK's first Maritime Innovation Hub.[10]

The Port's strategic choices comprise more than 20 innovation projects – its innovation pipeline – across 7 programmes, which the Port of Tyne calls strategic themes. Each project is overseen by a leadership team sponsor, led by a project manager and delivered by a multifunctional team of employees in a wide variety of roles across the business. Every project has a 2050 vision and rolling innovation plans and targets which are formally updated annually, with other changes processed as needed throughout the year. Progress against these plans is reported monthly to the leadership team and board. Updates are also communicated to all employees across the business via Portall, the Port of Tyne's internal communications app.

Port of Tyne hosts regular stakeholder briefings to share progress updates on its Tyne 2050 sustainable innovation projects and frequently communicates specific project news and achievements through the press and its social media channels.

Sustainable innovation process

At its core, the sustainable innovation process is a mechanism that helps organizations to progress the most promising project ideas to becoming successful sustainable innovations. It helps to streamline sustainable innovation and reduce risk, providing a process to prioritize, screen, select, eliminate, refine and test proposed innovative solutions.

The stages in the innovation process are often represented as a funnel (Figure 9.2), with projects moving from the ideas stage through to implementation if they survive the various stages in between.

There are six key stages in the sustainable innovation process:

1 Problem identification: This initial step is crucial as it sets the direction for the entire innovation journey. It involves recognizing a gap or a need that warrants attention and could benefit from a sustainable solution.
2 Opportunity recognition and ideation: This is the brainstorming phase where multiple ideas are generated, without filtering them for practicality. It's about quantity and creativity, opening up to as many options as possible.
3 Clarification: In this phase, the emphasis shifts to understanding the problem or opportunity in much more detail. This involves refining the pool of ideas by giving them more structure and context.
4 Idea selection and development: This stage is where the best ideas are chosen based on their relevance, feasibility and potential for

FIGURE 9.2 Sustainable innovation process

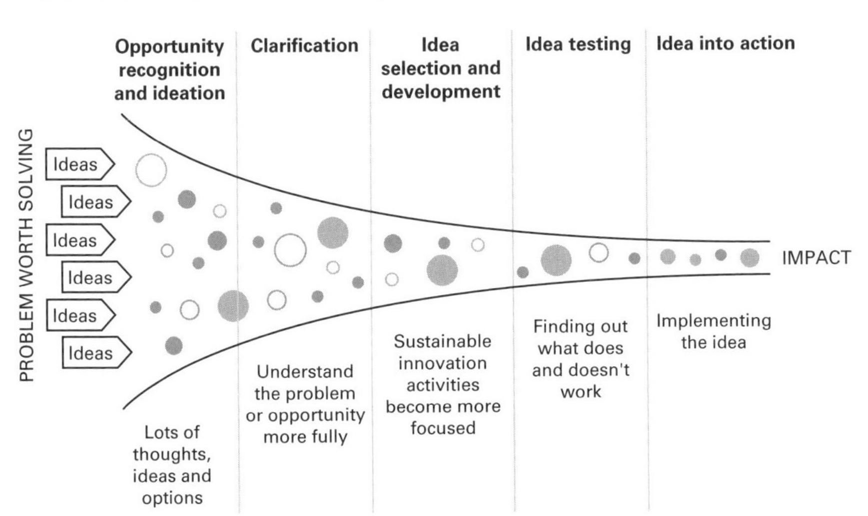

sustainable impact. These ideas are then fleshed out, developed further and prepared for testing.

5 Idea testing: This step involves experimenting to see what works and what doesn't. It's a critical evaluation phase where ideas are put to the test, often leading to further refinement.
6 Idea into action: The final phase is about implementation. It's where an idea, having passed through the previous stages, is executed to achieve the intended impact.

At the end of each stage, there is a decision 'gate', at which a decision-making team (often including senior management and project stakeholders) reviews the sustainable innovation's progress, evaluates the work completed, examines the latest findings and decides whether the project should proceed to the next stage. This series of reviews, known as the 'stage-gate process', is based on predefined criteria, which typically include aspects like technical performance, market viability, financial analysis and sustainability considerations. To ensure that well-informed decisions are made at each stage-gate, the sustainable innovation team needs to gather necessary data and research ahead of time so that decision-makers are equipped with comprehensive information on performance and other critical factors.

The stage-gate process provides a structured framework that helps organizations manage and mitigate risks by making informed decisions at critical points throughout the project lifecycle. It allows for the timely allocation of resources to projects with the highest potential and can be particularly effective in managing complex and high-risk projects.

In practice, progress on the sustainable innovation is usually not as linear as the process suggests. Insights gained from different stages often mean that the project development needs to go back a stage or two, in an iterative process.[11] The earlier stages of the process have an inherently higher level of uncertainty than the later stages. These focus on search and exploration, turning sustainable innovation ideas into proof of concept, then 'value propositions that matter to customers, embedded in scalable and profitable business models'.[12] The later stages are about execution of the ideas that have been developed and tested earlier.

Not all projects in the pipeline will survive the innovation process. The elimination effect is an important part of the process's purpose. Exiting a project as soon as it becomes clear that it's not viable is useful. It saves time and money and frees resource up to focus on more promising sustainable innovations.

On the other hand, once a project has been through all the stages in the innovation process, it should finally emerge as a tangible, desirable, feasible, viable and sustainable solution. The solution will have been developed and tested with the right amount of structure and governance, without being slowed down by unwieldy processes.

Combining the sustainable innovation process with lean innovation principles

Lean innovation is a methodology adapted from lean manufacturing principles, which originated in the Japanese automotive industry, notably with Toyota.[13] It emphasizes creating value for customers with fewer resources and minimizing waste – 'waste' being anything that doesn't add value to the customer. This approach focuses on understanding customer needs, rapid prototyping, continuous improvement and validated learning through customer feedback. Adapting these principles to innovation, lean innovation provides a framework for more effective and efficient innovation by creating minimum viable products (MVPs), testing them in the market and iterating quickly based on real-world feedback.[14]

When applied to long-term, complex, high-value, technical sustainable innovation projects, lean innovation principles can seem counterintuitive, due to the scale and complexity inherent in these projects. However, the core concepts of lean innovation – customer-centricity, waste minimization and continuous improvement – are universal and can bring some significant benefits, for example:

- Customer-centricity: for sustainable innovation projects, the 'customer' may include the end users, the operators or the broader community. Focusing on their needs ensures that the project delivers real value.

- Value stream mapping: by mapping out all the steps required to deliver the project, teams can identify and eliminate steps that do not add value, reducing costs, waste and time.
- Rapid prototyping and testing: even, and possibly especially, in large projects, designs, components or phases can be prototyped or piloted to test assumptions and gather feedback. This can prevent costly errors in later stages.
- Iterative development: complex projects can be broken down into smaller, manageable parts that can be completed in stages. This allows for iterative improvements and the flexibility to adapt to new information or changes in the environment.
- Feedback loops: establishing mechanisms to gather and act on feedback throughout the sustainable innovation project lifecycle can lead to continuous improvement and better alignment with stakeholder needs.
- Efficient resource management: lean innovation principles promote the efficient use of resources, which is critical in sustainable innovation projects where budgets and timelines are significant.

For core innovation projects, it is likely that the stages and requirements of the sustainable innovation process can be streamlined substantially. Conversely, transformational sustainable innovations may need a more specific and detailed approach, with different aspects of the innovation potentially needing their own sub-plan and route through the sustainable innovation process. However, it's important that your organization has a clear, well-understood sustainable innovation process that is adhered to.

It's also essential to measure the progress and performance of sustainable innovation projects as they move through the funnel, providing regular updates in the line with the governance processes set out in Chapter 8. This can be done through several, specific KPIs (key performance indicators) and metrics that are complementary to those KPIs and OKRs (objectives and key results) defined for the project following completion (Chapter 2). Here are some examples:

- Sustainability indicators: measure the environmental impact of the project, such as carbon footprint reduction, energy efficiency gains, waste reduction, water conservation or other relevant metrics.
- Financial metrics: track the cost savings or revenue generation against projections, return on investment (ROI) and payback periods to evaluate financial performance.
- Innovation milestones: monitor the achievement of key milestones in the innovation process, for example completing feasibility studies, prototypes, pilot tests and market launches.
- Time to market (TTM): track the time taken from ideation to market introduction. This can be an indicator of the efficiency of the innovation process and the organization's agility.
- Market adoption rate: once the sustainable innovation is launched, monitor its uptake in the market. This includes growth, market share and customer acquisition rates.
- Stakeholder engagement: evaluate the level of engagement and feedback from customers, employees and partners throughout the innovation process.
- Risk management: assess how effectively risks are being identified, mitigated and managed throughout the innovation stages.
- Learning and adaptation: gauge the organization's ability to learn from successes and failures and adapt its processes accordingly.
- Intellectual property: track the creation of intellectual property, such as patents filed or granted, which can be an indicator of innovation.
- Employee involvement: measure employee involvement in sustainable innovation through participation rates in ideation sessions, the number of suggestions submitted or the percentage of staff working on innovation projects.

- Project scalability: assess the potential for scaling each project, considering the ease of expansion, adaptability to different markets and ability to increase production or distribution.
- Regulatory and standards compliance: ensure the project meets relevant industry standards and regulatory requirements for sustainability.

Of course, you are unlikely to want or be able to measure all of these and will need to tailor these metrics to the specific context of the project and the strategic objectives of your organization. Regular reporting against these types of KPIs ensures that stakeholders are informed about the project's progress and that decision-makers can take corrective actions when necessary.

If you have a high volume of sustainable innovation projects moving through the process, it's also advisable to report on the number and type of projects at each stage. Things to share include how many projects have or have not survived each stage-gate, and why, and the number and quality of new projects entering the process. This data will give you a view of how your sustainable innovation pipeline is progressing overall.

Barriers to the sustainable innovation process

It's essential to understand the barriers to sustainable innovation in your industry, location and organization, so that you can build strategies to overcome or leverage these respectively in your organizational and team innovation process. Some can also be progressed by working in partnership with participants in your wider sustainable innovation ecosystem. The most cited barriers to sustainable innovation progress are time, money, regulatory constraints and lack of know-how.[15,16,17]

Overcoming time constraints

Time constraints can arise from the need for rapid innovation cycles, the slow adoption of new technologies or the urgency to meet sustainability targets. Often organizations find it challenging to afford the capacity to juggle doing their day job alongside progressing important but less urgent sustainable innovation projects. Potential solutions to overcoming the barrier of time include:

- Implementing agile project management techniques that can help streamline the development process, allowing for quicker iterations and adjustments based on feedback and outcomes. Agile methods are addressed in more detail in Chapter 12.
- Using strategic planning to prioritize projects, as discussed above, to concentrate efforts where they can achieve the most significant and timely effects.
- Employing rapid prototyping tools and techniques to test ideas quickly and refine them before full-scale development, saving time in the sustainable innovation cycle.
- Adopting lean principles and lean start-up methodologies to create minimum viable products (MVPs) to test market hypotheses with minimal resources and time, then iterating based on user feedback.
- Engaging in partnerships with other companies, research institutions and start-ups to provide access to additional resources, knowledge and technologies.
- Using open innovation platforms to tap into external ideas, technologies and solutions that can speed up the development of sustainable innovations.
- Giving teams the autonomy to make decisions and act quickly can reduce bottlenecks and speed up the sustainable innovation process.
- Ensuring that employees have the skills and knowledge to work efficiently and innovate effectively can reduce the time required to bring sustainable solutions to market.
- Using digital tools, such as artificial intelligence (AI), automation and the Internet of Things (IoT), can streamline operations, enhance decision-making and automate repetitive tasks, freeing up time to focus on sustainable innovation.[18]

Overcoming financial constraints

Lack of funding can sometimes be resolved by:

- Seeking alternative funding, taking advantage of government grants and subsidies designed to support sustainable innovation projects. Regularly monitor government announcements and industry news for new incentives related to sustainability. Be proactive in applying early for available grants, subsidies and tax incentives to secure funding before resources are depleted.
- Collaborating with a customer or key user to co-fund selected sustainable innovation projects, potentially also sharing any resulting intellectual property (IP) rights.

Dealing with regulatory constraints

Regulatory hurdles can take time to overcome and require the organization to engage with policymakers and to actively participate in policy discussions to shape regulations that support sustainable innovation. Stay ahead of the curve by thoroughly understanding your organization's regulatory landscape and the regulations relevant to your sustainable innovation projects. Engaging with legal experts and regulatory consultants can help you to remain compliant and, anticipate and prepare for upcoming changes.

Building strong relationships with regulators is also helpful. Open lines of communication can support mutual understanding and trust. Approaching regulators with a collaborative mindset is also helpful. Show how your sustainable innovation is designed to benefit your stakeholders and the community in alignment with public policy goals, such as environmental protection or economic development.

Look for regulatory 'sandboxes'. These are safe spaces where organizations can test their innovations at an early stage, without getting too bogged down in regulatory red tape. They also give you access to regulatory expertise and support. For example, the Information Commissioner's Office in the UK hosts free regulatory sandboxes for organizations innovating products and services that utilize personal data.[19]

Leveraging public opinion can also play a role in overcoming regulatory barriers. Public support for your sustainable innovation projects can lead to public policy and regulatory changes. For example, in 2023, California Governor Gavin Newsom proposed new measures to streamline the construction of clean infrastructure. This was in response to a concerted push by the public and expert advisors to accelerate the state's climate goals and job creation. The proposals aimed to cut project timelines by more than three years, save hundreds of millions of dollars and significantly reduce paperwork. An executive order was signed to set up a strike team to expedite clean infrastructure projects, employing an all-of-government strategy for planning and development.[20]

Participating in industry associations and groups can accelerate progress through lobbying and influencing with the power of a combined voice. For example, the British Ports Association represents the interests of UK ports to the various tiers of national and devolved governments in the UK and internationally. It has over 100 members, covering more than 400 ports, terminal operators and port facilities.[21]

Building know-how

Look for partnership opportunities with organizations that share similar sustainability goals to combine resources and expertise. Participate in industry groups focused on sustainability to benefit from shared learning, standards development and collaborative projects. Source consultancy expertise or identify new roles for recruitment to plug important specific skills and experience gaps in your organization.

Invest in continuous learning and development and use the strategies to develop your team and build a thriving sustainable innovation ecosystem for your organization that are detailed in Chapters 10 and 11.

Conclusion

Create a sustainable innovation strategy for your organization by identifying your strategic choices, deciding what and what not to focus on and understanding how your strategic choices interrelate and impact each other. Make sure that you have the right ratio of core, adjacent and transformational strategic choices for your specific organization. Convert these choices into a prioritized sustainable innovation pipeline of projects. Measure and track the progress of these projects, doing your best to remove or mitigate any known barriers to sustainable innovation, and use stage-gates to determine, based on good information, which projects should continue to receive investment in money and resources. Also report on the overall progress and health of the sustainable innovation pipeline, including how many projects survive or don't survive each stage-gate, and why.

ACTION CHECKLIST

- Develop your organization's sustainable innovation strategy. This includes identifying strategic choices and understanding how these strategic choices relate to each other.
- Make sure that the balance between core, adjacent and transformational strategic choices is appropriate for your organization.
- Convert your strategic choices into a prioritized pipeline of projects that you will deliver using your version of the sustainable innovation process, with its stage-gates, measurement, tracking and reporting.
- Use lean innovation principles throughout your sustainable innovation process to accelerate progress, minimize waste and optimize stakeholder value.
- Identify and remove or mitigate as many significant barriers to sustainable innovation for your organization as you can.
- Constantly update your sustainable innovation strategy and pipeline with any significant changes and ensure that your organization has the capability to respond to those changes with agility.

Notes

1. Porter, M E. What is strategy?, *Harvard Business Review*, November–December 1996, 61–78
2. North, J (n.d.) How to do competitor analysis, https://bigbangpartnership.co.uk/how-to-do-competitor-analysis/ (archived at https://perma.cc/5YRB-V4B6)
3. Nagji, B and Tuff, G (2012) Managing Your Innovation Portfolio, *Harvard Business Review*, https://hbr.org/2012/05/managing-your-innovation-portfolio (archived at https://perma.cc/DMB2-NNDJ)
4. O'Reilly, C A and Tushman, M L. Organizational ambidexterity: Past, present, and future, *Academy of Management Perspectives*, 2013, 27 (4), 324–38
5. Nagji, B and Tuff, G. Managing your innovation portfolio, *Harvard Business Review*, 2012, 90 (5), 66–74
6. Anthony, S D, Viguerie, S P, Schwartz, E I and Landry, S. The transformative business model, *Harvard Business Review*, 2015, 93 (10), 91–98
7. IDEO (n.d.) Design thinking, https://designthinking.ideo.com/ (archived at https://perma.cc/P7YU-EQN8)
8. Cooperleaf (2021) Northumbrian Water selects Copperleaf portfolio to streamline asset investment planning and management, www.copperleaf.com/news/northumbrian-water-selects-copperleaf-portfolio-to-streamline-asset-investment-planning-and-management/ (archived at https://perma.cc/DHL9-BRFV)
9. Cooperleaf (n.d.) Make high-value decisions to drive innovation for future generations, www.copperleaf.com/solutions-for-industry/water-wastewater-infrastructure-asset-management/ (archived at https://perma.cc/4XN4-8DAG)
10. Port of Tyne (2023) Tyne 2050, www.portoftyne.co.uk/news-and-media/publications/tyne-2050 (archived at https://perma.cc/G5RM-3C2Z)
11. North, J (n.d.) How to build an innovation funnel for business growth, https://bigbangpartnership.co.uk/how-to-build-an-innovation-funnel-for-business-growth/ (archived at https://perma.cc/6S3P-UK8M)
12. Osterwalder, A, Pigneur, Y, Smith, A and Etiemble, F (2020) *The Invincible Company: How to constantly reinvent your organization with inspiration from the world's best business models* (The Strategyzer Series), 1st edn., Wiley
13. Womack, J P, Jones, D T and Roos, D (1990) *The Machine That Changed the World*, Rawson Associates, New York
14. Ries, E (2011) *The Lean Startup: How today's entrepreneurs use continuous innovation to create radically successful businesses*, Crown Business, New York
15. Weidner, K, Nakata, C and Zhu, Z. Sustainable innovation and the triple bottom-line: a market-based capabilities and stakeholder perspective, *Journal of Marketing Theory and Practice*, 2020, https://doi.org/10.1080/10696679.2020.1798253 (archived at https://perma.cc/EQ5D-J7BD)

16 Bocken, N M P and Geradts, T H J. Barriers and drivers to sustainable business model innovation: Organization design and dynamic capabilities, *Long Range Planning*, 2020, 53 (4), 101950

17 Ahmed, A M, Sayed, W, Asran, A and Nosier, I. Identifying barriers to the implementation and development of sustainable construction, *International Journal of Construction Management*, 2021, 23 (8), 1277–88

18 Broccardo, L, Zicari, A, Jabeen, F and Bhatti, Z A. How digitalization supports a sustainable business model: A literature review, *Technological Forecasting and Social Change*, 2023, 187, 122146

19 Information Commissioner's Office (n.d.) The guide to the Sandbox, https://ico.org.uk/for-organisations/advice-and-services/regulatory-sandbox/the-guide-to-the-sandbox/ (archived at https://perma.cc/34BW-GH8M)

20 Office of Governor Gavin Newsom. Governor Newsom unveils new proposals to build California's clean future, Faster, 19 May 2023, www.gov.ca.gov/2023/05/19/governor-newsom-unveils-new-proposals-to-build-californias-clean-future-faster/ (archived at https://perma.cc/MP82-BD3S)

21 British Ports Association (2024) www.britishports.org.uk/ (archived at https://perma.cc/GC67-6EQW)

10

Creating a sustainable innovation ecosystem

CHAPTER OVERVIEW

Chapter 10 focuses on the importance of creating a sustainable innovation ecosystem. Having a sustainable innovation ecosystem is critical for fostering collaboration, knowledge exchange and cross-industry partnerships. This chapter explores strategies for building an enabling environment that supports sustainable innovation. From developing systems and policies to engaging diverse stakeholders, you'll gain insights into how to establish a robust ecosystem for your own organization. Themes also include successful collaboration approaches for working with a complex range of supply chain partners and creating an enabling environment through systems, incentives and supportive policies.

What is a sustainable innovation ecosystem?

As explored in Chapter 3, leaders of sustainable innovation face big challenges and opportunities in this historic period of climate change, technology advances, decarbonization and clear energy transition. Many of these challenges and opportunities are bigger than any single organization, entity or even industry or profession can solve alone. Creating and participating in a thriving sustainable innovation ecosystem is essential for organizations to survive, grow and deliver value in the current context of our fast-changing world.

The definition of a sustainable innovation ecosystem is a community in which businesses large and small, including start-ups, public sector organizations, academia and the not-for-profit, media and culture sectors, work together to shape new, purposeful solutions to challenges and opportunities.[1] These ecosystem partners from across the private sector, the public sector, research institutions and civil society are known as the 'quadruple helix' or the 'quintuple helix' when the ecosystem includes stakeholders representing the natural environment.

The concept of an 'innovation ecosystem' draws its metaphorical underpinnings from natural ecosystems. A natural ecosystem describes the interactions and relationships within a community of organisms and their environment. This concept has been adapted to describe innovation ecosystems, which refer to interconnected networks of entities such as companies, knowledge institutions and governments. These entities co-evolve around shared technologies, knowledge or skills to develop new products and services. Well-known and long-standing US examples are Silicon Valley for technology innovation and Hollywood for film and television. In sustainable innovation, ecosystems have emerged around renewable energy, sustainable agriculture, green construction, among many others. The ecosystem metaphor highlights key aspects such as cooperation, competition and symbiosis, which are crucial in both natural and sustainable innovation ecosystems.[2,3]

Sustainable innovation ecosystems can be large or small, co-located physically or virtually. The participating organizations have multiple direct and indirect links between them. Ideas, knowledge, connections, finance and people resources cycle between participants, as successful organizations reinvest back into the innovation ecosystem.

All organizations are a series of connected ecosystems. However, sustainable innovation ecosystems often have their own identities and stated purpose and carry out activities that are targeted to achieve sustainable innovation.

Mutually beneficial success

Every organization in a sustainable innovation ecosystem depends on the success of the other organizations, either directly or indirectly.

The successes of individual organizations can have far-reaching positive impacts, setting into motion a virtuous cycle of mutual benefit. When one entity innovates, it generates a wealth of knowledge that can permeate through the ecosystem, elevating others and providing a knowledge base that can be expanded more widely. This dissemination of knowledge, whether through academic publications, industry conferences or casual exchanges, creates a catalyst for collective growth.

As organizations progress their sustainable practices, their supply chains also adapt and evolve, shaping a network of efficiency and sustainability that benefits all links in the chain. This transformation can lead to improved sustainability benefits, production methods, cost savings and an overall increase in the quality of goods and services provided. The collective reputation of the ecosystem also gets a boost from these individual successes, attracting a higher calibre of talent, partners and customers all eager to be part of a forward-thinking and prosperous community.

Regulatory advancements often follow innovation. If one organization's sustainable practices influence policy changes, such as incentives for green technology, this can open doors for others in the ecosystem to step through and benefit from them, too. The network effect also comes into play. The greater the number of organizations that join and engage with the ecosystem, the more valuable the network becomes for everyone. This proliferation of resources, expertise and collaborative opportunities creates fertile ground from which new ideas can sprout and grow.

CASE STUDY

Autobahn A3 project – Germany's largest PPP infrastructure project

A public–private partnership (PPP) is a cooperative arrangement between one or more public and private sectors. Typically used in infrastructure projects or public services, PPPs involve private entities financing, designing, implementing and operating projects that traditionally have been provided by the government. This collaboration allows for shared risks and resources, often leading to increased efficiency, innovation and the ability to complete large-scale projects more effectively.

The Autobahn A3 project in Germany is a prime example of a successful PPP, representing the country's largest PPP infrastructure endeavour, and illustrates a sustainable innovation ecosystem. This ambitious project involves the expansion of a 76-km section of the Autobahn A3 to six lanes, aiming to enhance traffic flow, safety and environmental sustainability. Managed by the Autobahn GmbH of the Federal Northern Bavaria and executed by A3 Nordbayern GmbH & Co. KG, the project combines government oversight with private sector efficiency and innovation.

This partnership enables rapid implementation and incorporates advanced digital tools like 5D Building Information Modelling (BIM) for transparency and efficiency. The project also encompasses long-term maintenance and operations, ensuring quality and minimal traffic disruptions over a 30-year period. Construction started in May 2020 and is scheduled for completion by the end of 2025.[4,5]

Open innovation

Open innovation is an approach taken by an organization or network to access the ideas, technology and knowledge that are available externally, beyond its employees and existing supply chain.

It develops when participants in sustainable innovation ecosystems expand their organizational resources and seek new ideas, knowledge and solutions from external collaborators. This promotes the flow, clustering and bringing together of resources within a sustainable innovation ecosystem.[6]

Open innovation started in the commercial sector, with businesses such as IBM, Unilever and Phillips taking the lead. It is now becoming more used in the social sector, and by government organizations.

For example, organizations such as the US Space agency, NASA, publish innovation challenges and crowdsource solutions for them. Their open innovation programme includes student and citizen science initiatives; aeronautics; astronomy; earth science; engineering and modelling; health and medicine; IT; living in space; multimedia production and design; physical sciences; and planetary science. They run competitions and challenges with prizes.[7]

Sticking solely to internal R&D can limit an organization's sustainable innovation potential. The open innovation model allows organizations to tap into external knowledge and resources and collaborate with external partners to generate the best solutions for mutual benefit. There are numerous ways to achieve open innovation. Organizations can participate in joint ventures, become part of an open innovation platform or collaborate with research institutions. Recent years have also seen an increase in open business models that leverage external resources for successful sustainable innovation. Open innovation is a crucial part of building a thriving sustainable innovation ecosystem. It encourages the exchange of ideas, resources and technologies beyond the boundaries of any single organization.

CASE STUDY

National Nuclear Laboratory and open innovation

The National Nuclear Laboratory (NNL) is owned and operated by the UK government and plays a pivotal role in the nuclear industry. Its primary function is to provide expertise in nuclear science and technology, serving as a bridge between the public and private sectors. NNL operates as a centre of excellence, conducting vital research and development to advance nuclear fuels, materials and processes.

NNL's responsibilities include supporting the UK's existing nuclear power stations, aiding in the decommissioning of legacy sites and developing technology for new nuclear build projects. It also plays a key role in non-proliferation through material safeguards and helps to ensure the UK's energy security and environmental objectives are met.

As an advisor to the UK government, NNL contributes to shaping policies that guide the future of the country's energy landscape. Additionally, it fosters innovation in the nuclear industry by collaborating with universities, research institutes and other organizations worldwide. Through these efforts, NNL aims to maintain and grow the UK's nuclear capabilities, ensuring safe, clean and efficient energy production for the future.

Dr Tim Whitworth and Paul Knight, NNL's head of innovation and technology commercialization manager, respectively, founded NNL's Open Nuclear Innovation Platform, driven by a recognized need for a more open approach to sustainable innovation within the nuclear sector across the four strategic areas of health and nuclear medicine, clean energy, environmental restoration, security and non-proliferation.

Tim and Paul launched the Open Innovation Platform through their already extensive list of contacts. They leveraged social media and webinars for promotion and pinpointed specific companies and institutions that could potentially offer solutions to NNL's specific innovation challenges.

Challenge-led open innovation is a process where specific problems or 'challenges' are presented as open calls for solutions to a broad audience, often transcending organizational boundaries. This approach hinges on the principle that the best ideas may come from outside the traditional confines of the organization. The benefits of this method are many. It leverages diversity, bringing in a range of perspectives and disciplines to bear on a problem. It accelerates innovation by connecting problem-solvers with those who need solutions. In addition, it democratizes the innovation process, as anyone with a viable idea can contribute.

Tim and Paul work closely with colleagues across NNL to identify appropriate challenges for the Open Nuclear Platform. Tim explains:

> In several projects, we've encountered situations where the initial understanding of a challenge evolved over time. We've held meetings to define a challenge, developed solutions and presented them, only to find out later that the requirements had changed. This highlights the importance of having a crystal-clear understanding of the challenge from the outset, especially when working with external partners.
>
> Being precise about the expected outcomes is crucial, as is preventing a change in ideas or expectations over the course of the project. To mitigate this, establishing a consistent feedback loop is essential. This ensures that any shift in objectives or vision is communicated promptly, allowing the development team to adjust their trajectory accordingly. Understanding the timeline and being transparent about the process with all stakeholders involved are also critical factors for the success of a project.

Examples of challenges include developing long-lasting seals for nuclear product cans,[8] remote characterization of mixed, contaminated waste,[9] and the export and handling of small, irradiated samples.[10] Organizations with potential solutions to these challenges can seek funding through the platform. Those selected receive up to £12,000 to further investigate their ideas in a 'discover' project. The funding's goal is to assist organizations in collaborating with NNL to determine the feasibility and potential of their proposals. Following the 'discover' funding stage, organizations can advance their solution in collaboration with NNL.

Paul explains how NNL helps these organizations to advance their potential challenge solutions through the Technology Readiness Levels (TRLs),[11] using NNL's solid understanding of the development pipeline. He gives the following example:

> There is a team that specializes in handling unfamiliar materials. They have a systematic approach for identification, characterization, and determining treatment routes. They can test potential solutions on a small scale, and if successful, they scale up. If not, they design custom treatment routes. They're adept at scaling from a beaker to a 500-litre drum, understanding each step of the process. The combined expertise of the participating organization and NNL create a win-win for both parties.

The Open Nuclear Innovation Platform represents a significant development for NNL and the entire nuclear industry. It promotes innovation and cooperation, aiming to unearth new and improved methods to tackle technical challenges, and demonstrating NNL's commitment to cultivating a thriving sustainable innovation ecosystem, offering organizations the chance to contribute to and benefit from the industry's ongoing success.[12,13,14,15]

CASE STUDY

Sellafield's Game Changers innovation programme

Sellafield's nuclear decommissioning programme is a huge, ongoing effort to safely manage and dismantle a legacy of nuclear activity spanning decades. Located in Cumbria, UK, Sellafield Ltd operates on behalf of the Nuclear Decommissioning Authority (NDA), which is responsible for cleaning up the UK's earliest nuclear sites safely, securely and cost-effectively. The decommissioning process at Sellafield involves an intricate set of operations, from the retrieval and processing of nuclear waste to the demolition of redundant buildings and the remediation of contaminated land.

The decommissioning challenge at Sellafield is unique due to the site's history as a former plutonium production plant for nuclear weapons and a nuclear fuel reprocessing facility. Sellafield houses a complex mix of old and new facilities, each with its own specific risks and waste management needs. The decommissioning programme is focused on reducing the site's hazard and risk profile, all while maintaining the highest safety and environmental protection standards.

Sellafield engages in continuous research and collaboration with various industry and academic partners to innovate and find more efficient ways to carry out decommissioning tasks. By utilizing cutting-edge technologies and methods, Sellafield aims not only to expedite the decommissioning process but also to serve as a model for similar operations worldwide.

Sellafield's Game Changers programme is a forward-thinking initiative that significantly impacts the nuclear decommissioning sector. Its primary aim is to source and develop innovative solutions to some of the most challenging problems faced in nuclear decommissioning, particularly at sites managed by the Nuclear Decommissioning Authority (NDA) in the United Kingdom, including Sellafield.

Game Changers acts as a bridge between the nuclear decommissioning sector and the wider innovation community. It invites thinkers, developers and entrepreneurs from various industries to propose and develop their ideas to tackle specific, pre-identified challenges within the decommissioning process. These challenges range from waste management and material treatment to safety enhancements and efficiency improvements.

Game Changers is critical to nuclear decommissioning because it accelerates the development of new technologies and approaches needed to safely dismantle and remediate nuclear facilities. This process historically has been slow and costly due to the complex nature of nuclear waste and the hazards involved. By fostering a collaborative environment where innovation is not just encouraged but strategically directed towards real-world challenges, the Game Changers programme helps to ensure that the decommissioning work at Sellafield and other NDA sites is performed more effectively, safely and sustainably.

The programme is a catalyst for sustainable innovation in the nuclear decommissioning field, providing not only the platform for ideation but also the support for bringing viable solutions to fruition. Through Game Changers, the decommissioning process benefits from cutting-edge research and technology, drawing from a diverse pool of knowledge and expertise across sectors.

Dr Frank Allison, founder and CEO of FIS 360, the organization that developed and delivers the Game Changers programme on behalf of Sellafield, says:

> Key to the success of the programme is the formulation of challenges in a way that resonates across industries. By stripping away the jargon and contextual complexities typically associated with the nuclear sector, Game Changers has been able to attract a broader range of problem-solvers.

Frank gives examples where the Game Changers programme has led to the development of breakthrough technology which could revolutionize the way in which stored waste is monitored, including sensors capable of operating for decades without battery replacement and remote hydrogen sensing technology that makes use of technology used in mobile phones and car radars. Game Changers has also facilitated the transfer of satellite technology, originally intended for outer space applications, to the medical field and finally, through adaptation, to nuclear applications. Frank says: 'This is a testament to the programme's ability to not only foster innovation but to adapt and apply it in various contexts for maximum impact.

He goes on to explain:

> The programme's methodology is highly structured, built on transparency and fairness. Potential solutions are meticulously evaluated against a set of criteria that includes impact, alignment with NDA's values and long-term viability. This rigorous process ensures that innovations are not only effective in addressing immediate challenges but are also sustainable and beneficial to the environment and society at large.

Frank's insights offer valuable guidance to professionals in the sustainable innovation space. He underscores the need for patience and a balanced expectation management, highlighting that cultural and procedural shifts within organizations take time to develop. Additionally, he emphasizes the importance of viewing sustainable innovation as an integral part of daily operations rather than an extracurricular activity.

The Game Changers programme demonstrates the benefits of developing a sustainable innovation ecosystem. Its success is a direct result of its ability to create a dynamic and responsive community that encourages the sharing of knowledge and expertise, ensuring that the challenges of today are met with the sustainable solutions of tomorrow.[16,17,18,19,20]

Why are sustainable innovation ecosystems important?

When they work well, sustainable innovation ecosystems bring benefits such as:

- A unified voice and influence for participants.
- Increased competitiveness, sector and/or place awareness, also leading to better economic, social and environmental outcomes (for example, through job creation and prosperity, and green growth).

- Improved effectiveness and efficiency, through combined activities and shared resources.
- More and stronger interpersonal networks.
- Greater shared and individual creativity, knowledge and skills development.

Sustainable innovation ecosystem policy

Because sustainable innovation ecosystems can generate significant benefits, some government organizations around the world actively support and promote their development.

EXAMPLES OF SUSTAINABLE INNOVATION ECOSYSTEM POLICY FROM AROUND THE WORLD

European Union (EU)

The EU states that it 'aims to create more connected and efficient innovation ecosystems to support the scaling of companies, encourage innovation and stimulate cooperation among national, regional and local innovation actors'.[21] The European Innovation Ecosystems (EIE) programme, part of the EU's wider Horizon Europe research and development funding activities, focuses on bringing together people or organizations whose goal is innovation. This includes the links between resources (such as funds, equipment and facilities), organizations (such as higher education institutions, research and technology organizations, companies, venture capitalists and financial intermediaries), investors and policymakers.[22]

United Kingdom

The UK Innovation Strategy asserts that the UK aims to learn 'from the pandemic to create the world's best innovation ecosystem'.[23] UK Research and Innovation (UKRI) is a non-departmental public body sponsored by the Department for Science, Innovation and Technology (DSIT). Its aim is to build a world-class research system, providing investment and support for researchers. The organization also facilitates collaboration and provides large-scale physical research facilities, equipment and knowledge resources for researchers to use.[24] Innovate UK operates under the UKRI umbrella, focusing on business innovation. Their relationship enhances the UK's research and innovation ecosystem by combining strategic direction with targeted support for businesses.[25]

Australia

Australia's National Innovation and Science Agenda includes innovation ecosystem principles as one of its top priorities: 'Working together: increasing collaboration between industry and researchers to find solutions to real world problems and to create jobs and growth.'[26] The government funds accelerator and incubator programmes for sectors such as digital and agritech.[27]

New Zealand

New Zealand has a thriving innovation ecosystem that encourages world-leading research and development (R&D) activity and resourceful ways of making innovation happen, especially through its dedicated government agency, Callaghan Innovation.[28,29]

Callaghan Innovation's team of more than 200 of New Zealand's leading scientists and engineers empower innovators by connecting people, opportunities and networks, and providing tailored technical solutions, skills and capability development programmes and grants co-funding. Their role is also to enhance the operation of New Zealand's sustainable innovation ecosystem, working closely with government partners, Crown Research Institutes, and other organizations that help increase business investment in R&D and innovation.

They have created Scale Up NZ, which they describe as 'a free platform... to navigate New Zealand's business and innovation ecosystem... to find and connect with collaborators and investors, track recent deals and investments, search for key players by sector or business stage, and explore up-to-date market data.'[30]

Singapore

The Singapore government takes what it describes as a 'sandbox approach' to innovation ecosystem development. This enables innovators, universities and companies to work together to innovate and test new products and ideas for the region and beyond.[31] Start-ups across all industries have access to incubator and accelerator programmes, with some tailored to specific sectors for more targeted support, tech advice and mentorship. These programmes also enable networking with peers in the same area, the Eduspaze for edtech and the Singapore Power Energy Advanced Research and Development (SPEAR) programme for Greentech, for example.

The country's legal and regulatory framework has made it a magnet for foreign start-ups, venture capital and corporate research activities. Its clear regulations on business setup, ownership, corporate governance, shareholder rights and bankruptcy are designed to strengthen the innovation ecosystem. Singapore also benefits from low corporate tax rates, with no capital gains tax, making it appealing to investors and venture capital funds.

Singapore has numerous technology research centres within its universities. These engage in advanced research with potential commercial applications, especially in greentech, supported by significant government funding and industry collaboration. For instance, the Energy Research Institute hosts a start-up incubator, using the government's support to translate research into viable business ventures.[32]

Sustainable innovation ecosystems either can have a formal structure, including a governance regime, or may be created and managed through more informal affiliation. At the heart of each thriving sustainable innovation ecosystem is a clear, relevant shared purpose. Sustainable innovation ecosystems often exist to solve shared challenges – such as sustainable economic growth, eco-innovation, and decarbonization, skills or materials shortages.

They also form to exploit shared opportunities, such as the development of a supply chain for a major manufacturer or sector.

CASE STUDY

Building a sustainable innovation ecosystem in the maritime sector

An example of building a thriving sustainable innovation ecosystem is the UK's Maritime 2050 Innovation Hub ('Innovation Hub') at Port of Tyne, in the UK. Partners in the Innovation Hub include the Department for Transport, Connected Places Catapult, Offshore Renewable Energy Catapult, Nissan, Drax, Accenture, Ubisoft, Royal Haskoning DHV, telecoms giant BT and consultancy Frazer-Nash, along with the Port of Tyne and several leading universities. The Innovation Hub operates as a dynamic catalyst, fostering collaboration among partners to advance the maritime sector and the broader logistics industry. Encouraging the exchange of ideas, harnessing cutting-edge research and development, promoting technological advancements and tackling shared challenges head-on, the Hub aligns seamlessly with the strategic vision outlined in the UK's Maritime 2050 Strategy.

When the Innovation Hub was launched in 2019, the UK's maritime industry, steeped in tradition and history, was being outpaced by more agile and innovative global competitors. The port itself, as we've seen in earlier chapters, used to be historically central to the UK's maritime strength but needed to pivot fast to become more relevant to changing markets.

The Innovation Hub became the vehicle for change, and not just for technology or process improvements; it was about creating an internal culture that welcomes new ideas, embraces change and encourages collaboration. The Innovation Hub's goal was to make innovation part of the port's DNA. Matt Beeton, who joined the Port as CEO in 2019, recalls: 'We knew we needed to do something different. The Innovation Hub gave us a platform to explore ideas – a focal point for our colleagues across the port to share their visions.'

Collaboration across boundaries

One of the Hub's first successes was breaking down silos within the port itself. The impact also expanded outward, engaging customers, competitors, regional organizations and maritime businesses around the world in a shared vision for a cleaner, smarter maritime future. 'It's about approach,' Matt explains. 'Being open to change, collaboration and bringing people together has a direct value for the port.' Matt continues: 'The Hub's efforts earned the port the reputation of being a convener of influence. Government entities and private sector partners began looking to the port as a willing and capable ally in their own innovation endeavours.'

Being a trust port[33] provided the necessary freedom to invest in long-term innovations without needing to pay out short-term shareholder dividends. This unique position allowed for a focus on industry-wide improvements and regional growth. 'We absolutely have to operate commercially, but at the same time it's about doing what's right for the region and the industry,' Matt comments. 'Our trust status gives us a safe space to innovate for the greater good.'

Measurable outcomes and strategic shifts

The Innovation Hub's impact is evident in several measurable outcomes. The port has been at the forefront of 5G technology, port decarbonization innovation, green shipping AI and VR initiatives. Rather than just being individual projects, these sustainable innovations are part of a comprehensive smart, green port strategy centred around data, which we explored in Chapter 6.

Matt explains:

> We have a lot of data. The Innovation Hub helped us to understand its value and use it as an encyclopaedia on how to run the port more efficiently. Through the Innovation Hub, we collaborated with world-class data scientists from the UK's National Innovation Centre for Data, based in Newcastle-upon-Tyne, and work openly with other ports through the Maritime Data Cluster to learn from each other.

The human element

Perhaps the most critical aspect of the Hub's success is the human element – building trusting relationships. The port's leadership recognized that trust is the cornerstone of any collaboration. 'Trust is what unlocks potential,' remarks Matt. 'The Hub has changed how we operate, not just with our colleagues but with the entire industry.'

Looking ahead

Looking ahead, the port plans to harness the momentum created by the Innovation Hub. 'We've only scratched the surface,' Matt says. 'The next phase is about amplifying this success and driving even greater innovation.'

The Innovation Hub at the Port of Tyne has demonstrated that fostering a sustainable innovation ecosystem requires more than just technology or process enhancements. It demands a cultural shift towards openness, collaboration and trust. The Hub has not only revitalized the port's operations but also positioned it as a regional and industry leader in sustainable innovation, ready to navigate the challenges of the future with confidence and creativity.

What sort of participants should a sustainable innovation ecosystem include?

A thriving sustainable innovation ecosystem needs to include a wide diversity of participants, from businesses at all stages, and all sizes, academia, government, not-for-profits, research organizations, funders and skills providers. Diversity matters because it creates greater richness of expertise, ideas, network connections and perspectives.

It is also important to include keystone organizations,[34] also known as platform leaders,[35] or ecosystem leaders,[36] in the ecosystem. These are the most significant members, acting as the anchors that ensure growth and stability in the ecosystem, supporting and connecting various elements. These organizations often lead in innovation and sustainability, influencing others through their practices, collaborations, ambition and example. Their functions include:

- Sustainable innovation leadership: fostering new ideas and technologies.
- Collaboration facilitation: connecting different entities such as start-ups, investors, and researchers.
- Resource sharing: providing access to critical resources like knowledge and funding.
- Standard setting: helping to establish best practices in sustainable innovation.
- Community building: creating a network of entities focused on sustainable innovation.

Keystone organizations create a ripple effect, where their actions and successes inspire and enable others to innovate and adopt sustainable innovation practices. The ecosystem also needs to include intermediaries, who play an important role in connecting different parties, facilitating iterative collaboration processes and helping to spread innovation, all of which are essential for building sustainable ecosystems and achieving Sustainable Development Goals.[37]

CASE STUDY

Connected Places Catapult: connecting people to spark sustainable innovation

The Connected Places Catapult (CPC) in the UK performs an important intermediary role for sustainable innovation ecosystems as the UK's accelerator for cities, transport and place leadership.

The organization provides impartial 'innovation as a service' for public bodies, businesses and infrastructure providers to catalyse step-change improvements in the way people live, work and travel. It connects businesses and public sector leaders to cutting-edge research to spark innovation and grow new markets. CPC runs technology demonstrators and small to medium-sized business accelerators to scale new solutions that drive growth, spread prosperity and eliminate carbon.

CPC is part of the Catapult Network, which comprises world-leading technology and innovation centres established by Innovate UK, the UK's national innovation agency.[38,39]

Challenges of operating in a sustainable innovation ecosystem

While there are substantial benefits of operating in a sustainable innovation ecosystem, there are also challenges:

- Participating businesses need to recognize when it is best to collaborate and when they will need to compete.
- Assessing how much time, resource and money should be invested, by whom and for whose benefit.
- Considerations around knowledge exchange and intellectual property protection. This is discussed in more detail in Chapter 11.
- Agreeing and managing governance and gaining consensus among participants on strategic and tactical decision-making.
- Practical considerations, such as finding times and locations when participants can be available to meet.
- Management of historic and ongoing rivalries and local organizational politics.

Sustainable innovation ecosystem success factors

Navigating these challenges effectively requires strong communication skills and a willingness to find common ground. This can be facilitated through regular meetings, joint projects or informal networking events, where different stakeholders can align their goals and share resources. Thriving sustainable innovation ecosystems indicate that the success factors are:

- Strong, accessible funding to accelerate high-potential innovations getting off the ground.
- The right mix of diverse ecosystem partners.
- A culture of collaboration to optimize ecosystem-wide ideation, knowledge exchange, cooperation and decision-making.
- A collective knowledge exchange platform.
- Innovation talent.
- Research and development (R&D) capability.
- Agility and responsiveness to external events, to pivot and innovate at speed in a fast-changing world.
- Effective keystone organizations that:
  - Facilitate the creation of a shared vision and value proposition.
  - Influence other potentially key players.
  - Make space and shape processes to cultivate sustainable innovation across the entire ecosystem.
  - Nurture mutually beneficial relationships between all types of participants.
  - Ensure that the sustainable innovation ecosystem adapts and evolves in line with external changes, to best support its participants.

Synergy between in-house innovation and external collaboration

For an organization to be in a strong position to benefit from the opportunities that come from being an active contributor to and participant in its own sustainable innovation ecosystem, its internal sustainable innovation teams need to have the necessary skills, capacity and mindset. They also need to be operating within an internal organizational culture that values and supports sustainable innovation.

Build a strong sustainable innovation team

Sustainable innovation teams are often seen as separate from the core organization, yet experience across various industries globally suggests there are benefits of establishing a dedicated group within an organization that concentrates exclusively on cultivating new ideas, solutions and products. This specialized team's primary objective should be to propel sustainable innovation and nurture creativity and progressive thinking to maintain the organization's competitiveness and significance.

While operating with some autonomy, it's essential that this team maintains a strong connection with the core business, adapting quickly in a rapidly evolving world to meet shifting needs and priorities. Integrating with the organization, grasping its essence, principles, purpose, mission, vision and values, is essential. This team should be empowered to venture beyond typical business limitations, to explore, test and determine the efficacy of diverse ideas.

To truly embed innovation into the fabric of an organization, including one that already prides itself on being innovative, requires a deliberate direction and unwavering commitment from leadership. Embracing sustainable innovation as a fundamental value and prioritizing continuous progress and evolution are vital. Emphasizing consistency, flexibility and a culture that prizes creativity, perpetual learning and the willingness to take calculated risks is key to driving sustainable growth.

Building a diverse team is essential for success. When individuals from various walks of life collaborate, they bring unique perspectives and ideas to the table. Effective communication and high levels of emotional and social intelligence make the diversity work.

While many people can acquire the skills necessary to be part of an innovation team, certain qualities set some individuals apart from others. It's possible to select team members who already possess these characteristics or develop them if the team member is interested and motivated to do so. In building a strong innovation team, leaders should look for:

- High integrity: people who are transparent and direct in their work, aligning with the organization's values and committed to implementing those values through sustainable innovation projects.

- Relevant technical expertise and experience.
- Collaboration: sustainable innovation often requires input from diverse stakeholders. It involves complex problems that span different fields and industries. Collaborating effectively allows innovators to pool expertise, share resources and integrate varied perspectives, leading to more robust, sustainable solutions.
- Networking: team members should have a strong network within and outside the organization.
- Self-efficacy: this means having the self-confidence to take a proactive approach when faced with a problem, having the ability to locate the resources necessary to resolve the issue and move forward.
- Grit: successful innovators persevere, attempting various paths to reach their goals and learn from their failures, becoming more determined to find a more appropriate solution.
- Excellence: both as a team player and as an independent thinker: having unique perspectives and ideas that they can integrate with others' ideas to create outstanding solutions. They can step back and look at the bigger picture, which enables them to adjust to the changes in the organization and the world around them. They keep their ideas fresh by thinking differently, laterally and connecting the dots.
- Market awareness: keeping up to date with the latest developments in the industry and demonstrating commercial acumen.
- Change management: the implementation of new ideas often meets internal resistance, making this an invaluable skill within the team.

Cultivating a culture of sustainable innovation

The activities and success of sustainable innovation teams are influenced to some extent by the perceptions of their organizational culture, and the organization's perceptions of the sustainable innovation team.[40] Organizational culture defines the environment in which teams operate. It sets the expectations, norms and values that guide behaviour within the organization. A culture that values sustainable innovation will likely encourage and support the sustainable innovation team, providing them with support, resources and recognition. This, in turn, can enhance the team's effectiveness and success.

Conversely, how the organization perceives its sustainable innovation team can affect the support and resources the team receives. If the organization sees the team as valuable and integral to its goals, it will likely invest more in its success. This perception can lead to increased trust, collaboration and openness to the sustainable innovation solutions the team proposes. How the leader of the sustainable innovation team engages and manages stakeholders can influence the organization's perception of the team. In essence, a positive, or negative, cycle of support and success can develop between the team and the organization, each reinforcing the value of the other. Leaders need to work to ensure that there is a positive relationship and alignment between the sustainable innovation team and the wider organization.

The organization's ability to absorb and implement new knowledge – its absorptive capacity – determines how well it can integrate internal and external innovations into its operations. It paves the way for smoother transitions and supports an environment that is receptive to sustainable innovation. A learning-oriented approach within the organization ensures that interactions with the internal sustainable innovation team, and external ecosystem, become opportunities for growth.

Building and nurturing relationships are vital in developing a strong network within the innovation ecosystem. These connections become the channels through which ideas flow and collaborations are forged. Additionally, the ability to assess and manage risks associated with external engagement ensures that the organization remains robust and secure in its core operations.

Using cross-functional collaboration to dismantle organizational silos will enrich the organization's internal sustainable innovation ecosystem. When teams from different functions unite for a common goal, it boosts engagement and a sense of ownership over the project. This leads to more commitment and drive towards achieving sustainable outcomes.

Leveraging sustainable innovation from the supply chain

The ability to realize sustainable innovation is what turns ideas into successfully delivered projects. It's this combination of skills and an innovative mindset that enables an organization to participate in an external innovation ecosystem, and thrive in it, and turn potential into tangible outputs and outcomes.

In terms of building a sustainable innovation ecosystem that goes beyond the organization, the supply chain is often a good and accessible place to start. Smaller supplier organizations may have agility, ideas and flexibility to offer; larger players could have resources, influence and R&D programmes that may be of benefit. Most suppliers will also have exposure to and insights from wider market developments, potentially also in adjacent sectors.

Before beginning to develop deeper strategic relationships with supply chain partners, you should assess their own effectiveness as a client. This means taking an honest view of how well an organization establishes an environment that encourages collaborative and high-performing teams together with suppliers, and its own behaviours as a client. These range from paying on time, transparent and reasonable contractual terms, through to being clear about expectations and working towards mutually beneficial and fair outcomes – essentially, being an organization that others want to collaborate with. If needed, get actionable, quantifiable feedback from your supply chain to help your organization to shape and implement an appropriate improvement plan. Carry out a comprehensive review of your organization's supply chain to establish which partners may be able to collaborate with you to achieve your sustainable innovation goals, in line with your organization's mission and vision. Actively engage with these companies and share your plans, hear theirs and be open to ideas.

If you find that your current supply chain does not have the capacity, capability or willingness to support your sustainable innovation ambitions, explore if and how these could be developed, potentially by working collaboratively to resolve any issues. Introducing several new, more innovative supply chain partners into your organization's sustainable innovation ecosystem to complement or replace some existing providers is also an option.

CASE STUDY

Equinor's ecosystem for sustainable innovation: focusing on suppliers, community and ecosystem

Equinor, headquartered in Norway, is a global energy company with a workforce of 22,000 across nearly 30 countries. Its goal is to lead in the energy transition. With a history of overcoming challenges in the North Sea for over 50 years, Equinor has grown into Europe's largest energy supplier and a leader in renewable energy and low-carbon solutions. It approaches challenges with a mindset that anything is possible until proven otherwise.

The company supports the Paris Agreement and the UN sustainability goals, striving to make energy both sustainable and accessible, while holding objective is to achieve net zero emissions by 2050. It acknowledges the energy transition as a critical global challenge and is dedicated to contributing solutions through knowledge sharing and collaboration.[41]

Leading sustainable innovation

Equinor is a major player in the UK energy sector, with a diverse portfolio that mirrors its global interests. It operates Mariner oil field and Hywind Scotland, the pioneering floating offshore wind farm. It's also expanding Norfolk's Sheringham Shoal and Dudgeon wind farms to power an extra 785,000 homes and exploring efficient offshore networks. Equinor leads H2H Saltend in Humberside for industrial CO2 reduction and collaborates on Scotland's Peterhead CCS Power Station, targeting 1.5 million tonnes of annual CO_2 capture.

World's largest wind farm

Equinor, alongside SSE Renewables and Vårgrønn, is developing the Dogger Bank UK offshore wind farm. Once finished, it will be the largest of its kind globally, more than doubling the size of today's biggest wind farm. Located about 130 km off the Yorkshire coast, it covers an area nearly as large as Greater London and almost twice that of New York City. First power was generated from Dogger Bank in 2023. Once fully operational, its 3.6 GW capacity from 277 offshore turbines will generate enough electricity to supply 6 million British homes every year. Each turbine stands at 260 metres tall, almost as tall as the Shard in London. These turbines will be installed and commissioned progressively, aiming for full commercial operation by 2026.

The company is harnessing the power of digitalization to ensure operational excellence. By collecting and analysing data from across the wind farm, the company can optimize maintenance schedules, improve turbine performance and enhance overall efficiency, ensuring that the wind farm operates with the smallest possible ecological footprint through its state-of-the-art control centre in South Tyneside. Equinor has collaborated with National Innovation Centre for Data in Newcastle to develop algorithms for its new system that will be used to manage all this data. Skills transfer has also been important here, with control room operators joining from a range of other industries, including the Tyne and Wear Metro, the local underground and overground light rail system. The control room has the capability to add future wind farms and is designed to be 'future-proof' and scaled up as required.

Equinor's approach to building the Dogger Bank wind farm includes a focus on creating a sustainable innovation ecosystem that comprises suppliers, employees, local communities and the surrounding environment.

Equinor's sustainable innovation ecosystem for Dogger Bank

The business has taken a proactive stance on developing sustainable innovation capability and capacity in its supply chain, engaging suppliers early on to ensure readiness for the project's ambitious scale. For example, the introduction of the GE Haliade-X 13 MW and 14 MW turbines requires suppliers to upgrade their capabilities, invest in new technologies and prepare for the future. This collaborative approach ensures that the innovations realized at Dogger Bank can be applied across the rest of the industry.

'Innovation excellence at Dogger Bank is not just about what we do, but who we do it with. Our strategic partnerships are the backbone of our operational capabilities,' says Tom Nightingale, Equinor's UK supply chain leader for renewables. 'We started with a huge emphasis on making sure that the supply chain knew what we were going to be doing and what was needed. The size and scale of the turbines at Dogger Bank mean that the supply chain have invested in new, specialized vessels just to install this size of equipment, setting a new industry standard for future offshore wind projects.'[42]

Innovation permeates the entire supply chain. From the turbines themselves to the supporting infrastructure, every component reflects a leap forward. Tom explains: 'We've also innovated by using HVDC in the UK for the first time for offshore wind and having unmanned platforms. It's not just about big turbine blades or the tallest turbines.' HVDC stands for High Voltage Direct Current. It's a technology used to transmit electricity over long distances by underwater cables in offshore wind farms, minimizing energy loss during transmission to the onshore grid.

Equinor collaborates with ports across the UK as part of its offshore wind operations, which is vital for the construction and maintenance of wind farms. It engages with a range of ports for various activities, including marshalling, installation and operational management. The ports facilitate the movement and assembly of large components, and their proximity to wind farm sites is crucial for efficient operations.

Equinor's commitment to sustainable innovation extends to operations and maintenance (O&M). The company's O&M strategy includes collaborating with its supplier North Star on the use of hybrid vessels, which, while currently reliant on diesel, have the capacity to switch to more sustainable fuels in the future. This forward-thinking design ensures that the fleet can adapt to advancements in fuel technology, reducing the carbon footprint over time. Equinor is also exploring infield charging for vessels, a move that could revolutionize energy use in offshore operations. North Star vessels also serve as operational bases for technicians, providing cutting-edge, cruise-ship standard facilities on board.

Built by predominantly local businesses, Equinor's O&M base in South Tyneside is the central hub for the wind farm. With a focus on being net zero, the offices include features such as solar panels, timber construction and facilities to support electric vehicles. This sustainable approach extends to the internal environment, featuring contemporary health and well-being facilities for employees.

Equinor's impact on local employment is significant. More than 2,000 jobs were created or supported by 2024, many concentrated in North East England. The project has injected new life into the local job market and accelerated the development of 'green' skills, ensuring that the workforce is future-ready. By providing training and development opportunities, Equinor ensures that the benefits of the Dogger Bank project will last for generations, contributing to a skilled workforce that can support the UK's green industrial revolution.

Training programmes at Equinor go beyond traditional methods. Technicians are multiskilled to handle high-voltage operations and turbine maintenance, creating a flexible and efficient workforce, meaning fewer specialists are needed offshore, thus reducing costs and simplifying logistics. 'Training our people to do those skills makes a lot of sense and it's a bit different from previous wind farm planning approaches,' says Tom.

Tom expands on this:

> Our investment in the local economy goes beyond job creation; it's about nurturing a community that grows with the green industrial revolution.

Equinor has been actively involving local communities, understanding their needs, and ensuring that the project's benefits are felt at a grassroots level. This includes educational initiatives, scholarships and grants that support local groups and STEM education, enabling the future of renewable energy and local development.

The investment in local people is evident in the £1 million dedicated to Equinor's community fund during construction, positively impacting people areas along the cable corridors and the operational base. Tom summarizes: 'Equinor's integrated approach ensures that every element, from the largest turbine to the smallest community initiative, contributes to sustainable innovation that benefits local people as well as accelerating the UK's journey to net zero.'

CASE STUDY

Northumbrian Water's Living Water Enterprise

Regulated by Ofwat, the AMP8 cycle (Asset Management Period 8) in the water industry in England and Wales, set to commence in 2025, presents a range of challenges that are complex and multifaceted, as discussed in earlier chapters. These challenges stem from various factors, including supply chain disruptions, energy price increases, skills shortages and sector-wide carbon targets.

One of the key aspects of AMP8 is the focus on environmental and regulatory challenges, with prominent themes such as achieving Net Zero by 2030, reducing sewage spills and continuing to mitigate the effects of climate change. The water industry is highly capital-intensive, and the supply chain is complex. There's a shift away from asset-intensive processes towards services such as nature-based solutions, digital technologies and demand management.

The relationship between regulators' requirements and inputs from the supply chain is indirect and complex, involving water companies' planning, procurement and the delivery of services to customers while protecting the environment. This calls for urgent and pragmatic decisions within the water industry to keep moving forward, with suppliers playing a crucial role.

The sector's response to these challenges will likely involve a combination of new technologies, innovative solutions and collaborative approaches. This includes optimizing ageing assets, adapting to climate change, recovering resources from wastewater and maximizing efficiency in biogas production. The use of digital twins and environmental modelling is also expected to play a significant role in developing sustainable solutions.

Northumbrian Water's Living Water Enterprise (LWE) is an industry-leading collaborative supply chain partnership, formed to deliver large-scale construction programmes, particularly for the upcoming AMP8 cycle and beyond. The LWE comprises Northumbrian Water and 12 supply chain partners who together will provide design and construction services.

The flexibility of the LWE allows these supplier partners to be available to work across Northumbrian Water's complete programme, ensuring the best resources are used for each project and focusing on long-term, smart spending and risk-based planning. The LWE is also focused on challenging existing methods and investment choices to explore alternative options that may include nature-based solutions for water and wastewater treatment, aiming for a reduction in carbon emissions.

This comprehensive initiative is Northumbrian Water's largest package of framework agreements and is intended to underpin the delivery of the company's substantial investment across its operating areas in North East England, Essex and Suffolk. The initiative is part of a strategic move by Northumbrian Water to build a culture centred around collaboration, sustainable innovation, safety and well-being, all while focusing on solutions that are beneficial for both customers and the environment. The partners selected for the LWE are expected to bring their energy and innovation to the projects ahead, contributing not just to the resilience of services for customers but also to building reputations and careers within the sector.[43,44,45]

How to build a sustainable innovation ecosystem for your organization and projects

To build a sustainable innovation ecosystem, start with these steps:

1. Create a clear purpose, vision and value proposition for the ecosystem, being mindful that it is likely to evolve as more partners come on board. You'll need to have some early ideas to hand, though, before you speak to potential partners.
2. Map out the first stage of the innovation ecosystem visually, including everyone you need on board in the early stages. Don't

include too many or too few people too soon, though. Make it manageable, with enough people to gain early traction. Create a sustainable innovation ecosystem map (Figure 10.1).

3 Speak to the members you think you must have on board from the outset to gain their buy-in and commitment.

4 Arrange your first innovation ecosystem session. This can be an innovation sprint[46] or workshop to collaborate on how you're going to proceed to create a successful, thriving innovation ecosystem.

FIGURE 10.1 Sustainable innovation ecosystem map

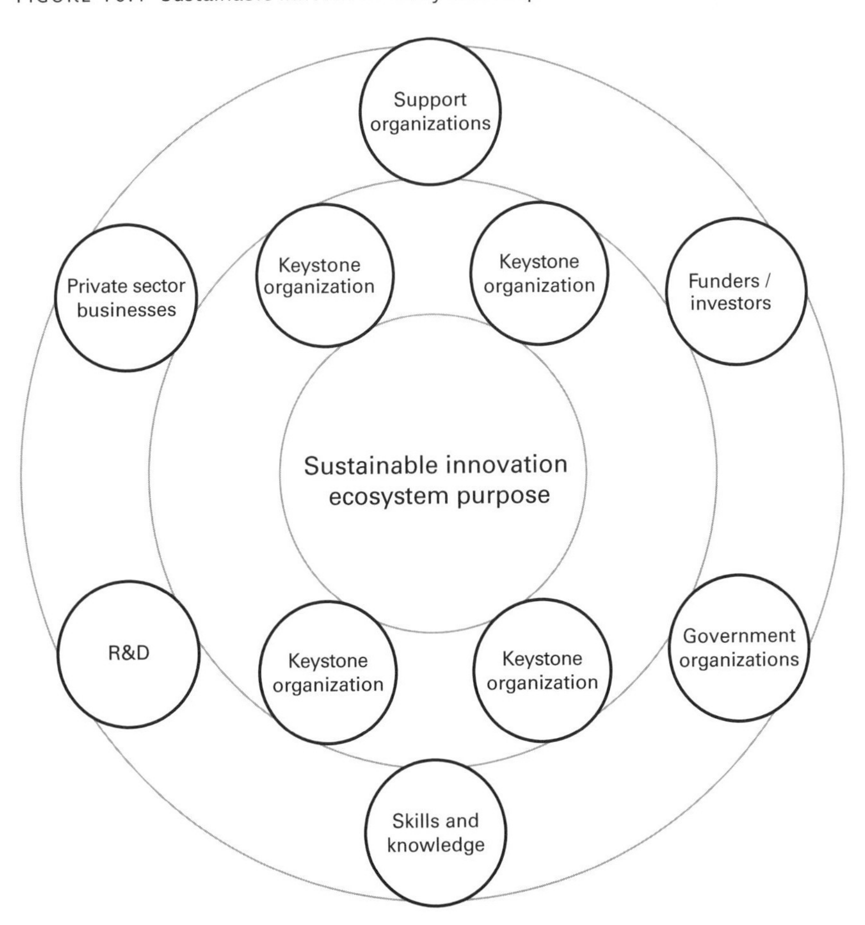

SUSTAINABLE INNOVATION ECOSYSTEM MAP

Because each sustainable innovation ecosystem is unique, comprising multiple participants with complex relationships, it is often useful to create a 'sustainable innovation ecosystem map'.

This is a visual representation of all the stakeholders directly and indirectly involved, showing their relative importance, connections and relationships with each other.

Use the map to pinpoint the sustainable innovation ecosystem's purpose and value proposition, how collaborative innovation is intended to add collective value to participants' success, the activities that will take to achieve them, the resources needed and potential risks and associated management actions.

Conclusion

'A rising tide lifts all boats' is an aphorism that applies to a thriving sustainable innovation ecosystem. Improved market conditions, problem-solving and solutions to shared challenges can benefit all participants, especially where the challenges are so significant that a single organization or a small number would struggle to tackle them alone. Most organizations benefit from cultivating and participating in their own innovation ecosystem, even though sometimes this requires additional effort to make things work. It is important to create a thriving internal organizational ecosystem, as well as an external one. Successful sustainable innovation projects demand a high-performing team, plus an organizational culture and leadership that understand the importance of sustainable innovation and actively supports it. The potential to leverage sustainable innovation via collaboration with supply chain partners should also be included in sustainable ecosystem development. Taking a strategic approach to cultivating a thriving sustainable innovation ecosystem optimizes the efficiency and effectiveness of an organization's engagement and collaboration activities.

ACTION CHECKLIST

- Carry out a strengths and gaps analysis of your sustainable innovation team and wider organizational culture. Identify actions to leverage the strengths and address any important gaps.
- Review the effectiveness of your efforts to collaborate with your supply chain partners on your sustainable innovation projects. Create a plan to achieve greater value and advance your sustainable innovation projects from your collaboration.
- Create a sustainable innovation ecosystem map for your organization and/or project. Work to fill any gaps and reach out to potential new partners or collaborators. Also reconnect with any lapsed partners and collaborators. Be intentionally and productively active within your sustainable innovation ecosystem.

Notes

1. North, J (n.d.) How to build a thriving innovation ecosystem, The Big Bang Partnership, bigbangpartnership.co.uk/how-to-build-a-thriving-innovation-ecosystem/ (archived at https://perma.cc/ZK3B-FTS8)
2. Moore, J F (1993) Predators and prey: A new ecology of competition, *Harvard Business Review*, 71 (3), 75–86
3. Osterwalder, A, Viki, T and Pigneur, Y. Why your organization needs an innovation ecosystem, *Harvard Business Review*, 15 November 2019, hbr.org/2019/11/why-your-organization-needs-an-innovation-ecosystem (archived at https://perma.cc/QK8A-FQ8D)
4. European Investment Bank (2020) Germany: EIB supports upgrade to A3 motorway in Bavaria, www.eib.org/en/press/all/2020-102-eib-supports-upgrade-to-a3-motorway-in-bavaria (archived at https://perma.cc/QZ97-RV3A)
5. Eiffage (2020) Eiffage and JOHANN BUNTE Bauunternehmung win the public-private partnership for the Biebelried-Fürth/Erlangen section of the A3 motorway in Germany, www.eiffage.com/news/eiffage-and-johann-bunte-bauunternehmung-win-the-public-private-partnership-for-the-biebelried-furtherlangen-section-of-the-a3-motorway-in-germany (archived at https://perma.cc/33G5-P4UZ)
6. North, J (n.d.) How Open Innovation can help your business, The Big Bang Partnership, bigbangpartnership.co.uk/how-open-innovation-can-help-your-business/ (archived at https://perma.cc/5QY8-R5R9)

7 National Aeronautics and Space Administration (n.d.) Open Innovation: Boosting NASA higher, faster, and farther, www3.nasa.gov/sites/default/files/atoms/files/fy1920openinnovationreport-final.pdf (archived at https://perma.cc/6HHQ-8PMK)

8 National Nuclear Laboratory (n.d.) Long-lasting seals for nuclear product cans, Open Nuclear, opennuclear.nnlinnovation.wazoku.com/challenge/07c28d5acc944666b966c0c5ceb9933f?entities=idea&sort=-relevancy&page=1&pageSize=15&parentType=challenge&parentId=07c28d5acc944666b966c0c5ceb9933f&communityId=7b13e7d5eeb14e099b7169e9fc27ea71 (archived at https://perma.cc/Y6GU-4UXL)

9 National Nuclear Laboratory (2023) Remote Characterisation of Mixed Contaminated Waste Challenge, October 2023, wazoku-clients.s3.eu-west-2.amazonaws.com/nnlinnovation.wazoku.com/Documents/ON_892_Red_Bunker_Challenge_Remote_Characterisation_of_Mixed_Contaminated_Waste.pdf (archived at https://perma.cc/4FPG-SR6S)

10 National Nuclear Laboratory (2023) Export and handling of small, irradiated samples, April 2023, wazoku-clients.s3.eu-west-2.amazonaws.com/nnlinnovation.wazoku.com/documents/2127f4e71e4a4754865b54331278710d.pdf (archived at https://perma.cc/Z52E-ZL95)

11 Nuclear Decommissioning Authority (2014) Guide to Technology Readiness Levels for the NDA Estate and its Supply Chain, assets.publishing.service.gov.uk/media/5a7f96f7e5274a2e8ab4d194/Guide-to-Technology-Readiness-Levels-for-the-NDA-Estate-and-its-Supply-Chain.pdf (archived at https://perma.cc/P8U7-3YYR)

12 National Nuclear Laboratory (n.d.) Open Nuclear Platform, www.nnl.co.uk/innovation-science-and-technology/showreel/collaborations/open-nuclear-platform/ (archived at https://perma.cc/L9MG-4CNJ)

13 Interview with the author, 15 November 2023

14 National Nuclear Laboratory (n.d.) NNL Innovation, opennuclear.nnlinnovation.wazoku.com/home-page (archived at https://perma.cc/5YRS-DYVK)

15 National Nuclear Laboratory (n.d.) www.nnl.co.uk/ (archived at https://perma.cc/K3RU-5KKW)

16 Interview with the author, 13 November 2023

17 Game Changers (n.d.) www.gamechangers.technology/ (archived at https://perma.cc/8LPH-P5MP)

18 Sellafield Ltd (n.d.) www.gov.uk/government/organisations/sellafield-ltd (archived at https://perma.cc/P5JB-9SYN)

19 Nuclear Decommissioning Authority (n.d.), www.gov.uk/government/organisations/nuclear-decommissioning-authority (archived at https://perma.cc/5C45-NVZD)

20 FIS 360 (2020) Game Changers, fis360.com/2020/11/10/game-changers-case-study/ (archived at https://perma.cc/HG23-FR5S)

21 European Commission (n.d.) Policy and strategy, research-and-innovation.ec.europa.eu/funding/funding-opportunities/funding-programmes-and-open-calls/horizon-europe/european-innovation-ecosystems_en (archived at https://perma.cc/AR8H-MCL2)

22 European Commission (n.d.) European Innovation Ecosystems, eismea.ec.europa.eu/programmes/european-innovation-ecosystems_en (archived at https://perma.cc/8ACF-HCYH)

23 Department for Business, Energy & Industrial Strategy (2021) UK Innovation Strategy: Leading the future by creating it, assets.publishing.service.gov.uk/government/uploads/system/uploads/attachment_data/file/1009577/uk-innovation-strategy.pdf (archived at perma.cc/M4MT-H47G)

24 UK Research and Innovation (n.d.) What we do, www.ukri.org/what-we-do/ (archived at https://perma.cc/EN9L-MG5A)

25 UK Research and Innovation (n.d.) About Innovate UK, www.ukri.org/who-we-are/innovate-uk (archived at https://perma.cc/9LJQ-G7XH)

26 Department of Industry, Science and Resources (n.d.) Technology, www.industry.gov.au/science-technology-and-innovation/technology (archived at https://perma.cc/3P78-ZABB)

27 Australia Trade and Investment Commission (n.d.) Explore innovations hubs, www.globalaustralia.gov.au/why-australia/innovation-australia/innovation-map (archived at https://perma.cc/HV2L-GZEW)

28 Ministry of Business, Innovation & Employment (n.d.) New Zealand research and development, www.mbie.govt.nz/science-and-technology/science-and-innovation/international-opportunities/new-zealand-r-d/ (archived at https://perma.cc/HU72-V5XP)

29 Callaghan Innovation (n.d.) About us, www.callaghaninnovation.govt.nz/about-us (archived at https://perma.cc/Y46M-JV4V)

30 Callaghan Innovation (n.d.) Scale-up NZ Connections Lab. www.callaghaninnovation.govt.nz/contact/scale-up-nz-connections-lab/ (archived at https://perma.cc/SB2D-QV3C)

31 EDB Singapore (n.d.) Innovate for business growth, www.edb.gov.sg/en/our-industries/innovation.html (archived at https://perma.cc/53FX-GK9W)

32 Development Asia (2023) https://development.asia/insight/why-singapores-ecosystem-technology-startups-among-best (archived at https://perma.cc/2SWS-YSBK)

33 Tyers, R and Brione, P (2022) Ports and shipping FAQs research briefing, House of Commons Library, p. 11, researchbriefings.files.parliament.uk/documents/CBP-9576/CBP-9576.pdf (archived at https://perma.cc/S724-6XXU)

34 Iansiti, M and Levien, R (2004) Strategy as ecology, *Harvard Business Review*, 82 (3), 68–78, 126. PMID: 15029791

35 Gawer, A and Cusumano, M A (2013) Industry platforms and ecosystem innovation, *Journal of Product Innovation Management*, doi.org/10.1111/jpim.12105 (archived at https://perma.cc/3EE9-57ZT)

36 Moore, J (1993) Predators and prey: A new ecology of competition, *Harvard Business Review*, hbr.org/1993/05/predators-and-prey-a-new-ecology-of-competition (archived at https://perma.cc/UU55-P8UL)

37 Sultana, N and Turkina, E (2023) Collaboration for sustainable innovation ecosystem: The role of intermediaries, *Sustainability*, 15 (10), 7754, doi.org/10.3390/su15107754 (archived at https://perma.cc/ZMX3-5LDJ)

38 Connected Places Catapult (n.d.) cp.catapult.org.uk/ (archived at https://perma.cc/YDC2-RY42)

39 UK Research and Innovation (n.d.) Innovate UK, www.ukri.org/councils/innovate-uk/ (archived at https://perma.cc/CL7K-E4SD)

40 North, J (2015) Individual intrapreneurship in organisations: A new measure of intrapreneurial outcomes, PhD thesis, University of York, core.ac.uk/download/pdf/42606125.pdf (archived at https://perma.cc/8Q2D-R8XY)

41 Equinor (n.d.) About us, www.equinor.com/about-us (archived at https://perma.cc/7TED-285T)

42 Interview with the author, 16 February 2023

43 Thompson, L (n.d.) Is the water sector ready for AMP8?, *Water Industry Journal*, www.waterindustryjournal.co.uk/is-the-water-sector-ready-for-amp8 (archived at https://perma.cc/WE7F-SX4Y)

44 Royal HaskoningDHV (n.d.) AMP8 – Facing up to future challenges, www.royalhaskoningdhv.com/en/markets/water/uk-water-utilities/amp8-facing-future-challenges (archived at https://perma.cc/3CTZ-GFAT)

45 Stantec (2023) Northumbrian Water appoints Stantec as strategic technical partner for AMP8, www.stantec.com/en/news/2023/northumbrian-water-appoints-stantec-strategic-technical-partner-amp8 (archived at https://perma.cc/XV3Y-M295)

46 North, J (n.d.) How to Build a Thriving Innovation Ecosystem, The Big Bang Partnership, bigbangpartnership.co.uk/how-to-build-a-thriving-innovation-ecosystem/ (archived at https://perma.cc/GJ8J-GV8M)

11

Knowledge exchange, ideation and problem-solving models for multidisciplinary expert teams

CHAPTER OVERVIEW

The purpose of Chapter 11 is to provide practical guidance and a toolkit for successful knowledge exchange, ideation and problem-solving models for multidisciplinary, expert teams. It includes:

- Understanding the importance of knowledge exchange, ideation and problem-solving in sustainable innovation.
- Practical intellectual property considerations when engaging in knowledge exchange activities with external partners.
- Demonstrating the importance of trust, psychological safety, communication, mutual respect and constructive challenge in developing high performing, multidisciplinary expert teams.
- Different options for facilitating knowledge exchange events to support your sustainable innovation ecosystem.
- Exploring innovation sprints, design sprints, concurrent design and charrettes as methods and toolkits to support an accelerate collaborative multidisciplinary sustainable innovation.

Getting the best from multidisciplinary expert teams

Multidisciplinary expert teams are groups of professionals with expertise in various disciplines who all collaborate towards a common goal. These teams are critical for sustainable innovation projects. They harness the unique skills, perspectives and knowledge areas of team members to address complex sustainable innovation problems and achieve practical, novel solutions. A multidisciplinary expert team can consist of people from the same organization, although, more commonly for significant sustainable innovation projects, it will comprise suppliers, advisors, other external partners and increasingly customers and key stakeholders.

Selecting the right team and creating a positive team and organizational culture are fundamental to sustainable innovation success, as we've seen in the previous chapter. However, to leverage the advantages of a robust, sustainable innovation team within a flourishing broader innovation ecosystem, along with fostering a positive culture of innovation within the organization, it's also essential to implement effective working practices and equip the team with the right support and collaboration skills.

The challenges of multidisciplinary expert team working

Even when everyone working in a multidisciplinary expert team is highly skilled, knowledgeable, collaborative and committed, progress can be slow and success limited, despite best endeavours. This is often due to one or more of a number of reasons.

Even with same shared mission, vision and values, team members are likely to have their own priorities and objectives. Ideally, this leads to positive, helpful creative tension, defined as respectful disagreement or discord about the best way to approach a problem or opportunity which ultimately, through discussion and debate, gives rise to better ideas or outcomes. The worst-case scenario, however, is that differences in priorities and objectives culminate in rifts, unhealthy disagreements and politicking. Different disciplines use their own vocabulary, or 'jargon', meaning that communication can be less effective.

Thinking style preference refers to an individual's favoured approach to processing information and making decisions. It includes the patterns of thought that shape how a person analyses situations, solves problems and generates ideas. These preferences can vary widely among professionals in multidisciplinary expert teams. Some may prefer data-driven, analytical thinking, while others might lean towards intuitive or creative approaches. Understanding thinking style preferences can help to optimize teamwork, as it allows for the alignment of tasks with individuals' strengths, promotes better communication and enhances problem-solving efficiency. But when these differences are not understood, it can lead to misunderstandings, frustration and even unhealthy conflict in teams.

Making progress as a multidisciplinary team can be slower and more time-consuming not only due to inherent differences within the team but also due to too much focus on achieving consensus. Building a culture where everyone signs up to 'disagree and commit' is much more effective. 'Disagree and commit' was used by Scott McNealy, co-founder of Sun Microsystems,[1] Andy Grove at Intel,[2] and later Jeff Bezos of Amazon.[3] Essentially it means that people in their team are encouraged to voice their perspectives and concerns freely. Then, after a decision is made, the commitment to execute the agreed-upon path is expected, regardless of initial disagreements. This ensures that the team can move forward together without prolonging unresolved debates. It facilitates rapid prototyping and experimentation and helps to maintain team unity and focus on shared goals, despite individual perspectives.

The importance of knowledge exchange, ideation and problem-solving in sustainable innovation

Successful multidisciplinary expert teams progress through the sustainable innovation process detailed in Chapter 9 by engaging effectively and efficiently in knowledge exchange, ideation and problem-solving together.

Knowledge exchange

Knowledge exchange is the dynamic process through which ideas, information and expertise are shared among individuals or organizations to mutually enhance understanding and capability. It is a collective endeavour that occurs within and across networks or partnerships, systematically and intentionally focusing on the co-creation, dissemination and application of knowledge.

Knowledge exchange is the foundation of the team's sustainable innovation process. It fuels sustainable innovation by facilitating the sharing of ideas, expertise and best practices, inspiring ideation and accelerating problem-solving by enabling cross-disciplinary and cross-industry learning.

There are numerous benefits of knowledge exchange for sustainable innovation:

- Organizations that actively engage in knowledge exchange are often perceived as sustainable innovation thought leaders and innovators in their fields. This can enhance their reputation, attract talent and improve their competitive position in the market.
- Learning from the successes and failures of others can help organizations to avoid common pitfalls, identify best practices and adapt more quickly to emerging sustainable innovation challenges and opportunities.
- Breaking down barriers between departments, teams and individuals allows for a more holistic and integrated view of the organization's knowledge.
- Tapping into external sources of knowledge, such as partners, customers and helps organizations stay up to date on the latest sustainable innovation trends, technologies and market demands, driving innovation and competitiveness.
- Facilitating networking and relationship building can lead to new partnerships, collaborations and business opportunities. These connections can provide access to additional resources, expertise and ideas that can spur sustainable innovation.
- Encouraging a culture of continuous learning and improvement assists organizations and individuals in constantly adapting, refining and evolving their ideas, products and processes.

A common question concerning knowledge exchange is around ownership and protection of intellectual property (IP). IP ownership in knowledge exchange can be complex because it involves the sharing and transfer of information, ideas and expertise between individuals, organizations or institutions.

Knowledge exchange can occur in various contexts, such as research collaborations, joint ventures, licensing agreements, consultancy services or educational programmes. In each scenario, IP ownership and rights need to be clearly defined and negotiated to ensure that all parties involved are protected and fairly compensated for their contributions.

IP AND KNOWLEDGE EXCHANGE FOR SUSTAINABLE INNOVATION

Here are some general principles and guidelines for IP ownership in knowledge exchange for sustainable innovation:

- Pre-existing IP:

  Before entering into any agreement or collaboration for sustainable innovation, it is important to identify and establish the ownership of any pre-existing IP. This includes patents, copyrights, trademarks and other intellectual property rights that each party brings to the collaboration. In most cases, pre-existing IP remains the property of the original owner unless explicitly agreed otherwise.

- Newly created IP:

  IP that is developed during the collaboration needs to be clearly defined and its ownership established. This may involve joint ownership, exclusive or non-exclusive licensing or assigning the IP rights to one party. The specific terms will depend on the nature of the collaboration, the contributions of each party to the sustainable innovation and any legal or contractual obligations.

- Confidentiality and non-disclosure agreements (NDAs):

  In many sustainable innovation projects, it may be necessary to share confidential information or trade secrets with other parties. To protect the IP rights and interests of each party, it is essential to have a confidentiality or non-disclosure agreement in place. This agreement should outline the specific information that is considered confidential, the obligations of each party to protect the information and any limitations or restrictions on the use and disclosure of the information.

- Publication and dissemination of results:

  Sustainable innovation involves the generation of new knowledge and the dissemination of this knowledge may be a key goal of the collaboration. However, it is important to balance the interests of all parties when it comes to publishing or sharing the results. This could involve negotiating terms regarding the timing and extent of publication, authorship and attribution and any restrictions or limitations on the use of the results for commercial or other purposes.

- Contractual agreements:

  To ensure clarity and protect the interests of all parties, it is crucial to have a written agreement in place that outlines the terms and conditions of knowledge exchange for the sustainable innovation, including the ownership and management of any IP rights. This agreement should be negotiated and agreed upon before the commencement of the project and should be reviewed and updated as needed throughout its duration.

IP ownership in sustainable innovation can be complex and requires careful consideration and negotiation to ensure that the rights and interests of all parties involved are protected. It is advisable to consult with legal experts and IP professionals when entering into any agreements or collaborations to ensure that the appropriate measures are in place to safeguard the IP rights of all parties.[4]

Ideation and problem-solving

Ideation simply means coming up with new ideas. The purpose of ideation is to devise novel concepts that could address specific challenges or capture opportunities with clear objectives in mind. Ideation is a critical stage in the sustainable innovation process, acting as the creative engine for generating, developing and refining ideas that have the potential to transform into innovative solutions. In this stage, the team often needs to think outside conventional paradigms to propose a variety of potential solutions, using creativity and divergent thinking to come up with a wide array of ideas, without immediate judgement or evaluation.

Ideation involves several key ingredients: problem identification to set the stage for ideation, inspiration to fuel creativity, team working to build on the ideas and structured techniques to guide the ideation towards actionable outcomes. It starts the problem-solving process that the team will need to engage in throughout the sustainable innovation process. Often, having a great idea is much easier than turning that idea into a reality. In ground-breaking sustainable innovation projects, the problems that multidisciplinary expert teams need to solve are either 'messy' or 'wicked':

- A messy problem is made up of clusters of interrelated or interdependent problems, or systems of problems. For example, the problems of decarbonizing a supply chain, the culture in a workplace or how to reach new markets sustainably are likely to be caused by multiple factors. The team will need to deconstruct messy problems and solve each key problem area.
- A wicked problem is a challenge that has unclear aims and solutions. Wicked problems are often challenges that keep changing and evolving. Some examples of current wicked problems are tackling climate change, obesity, hunger, poverty and more.

A few problems, of course, may be 'tame', i.e. they have a straightforward solution and can be solved through logic and existing know-how.

Collaboration: the foundation of successful ideation, knowledge exchange and problem-solving in multidisciplinary teams

Effective collaboration is at the heart of all successful knowledge exchange, ideation and problem-solving in multidisciplinary expert teams. It is a process in which people with diverse interests, skills and knowledge work together to leverage their collective wisdom, resources, insights, ideas and enthusiasm to improve outcomes and enhance decisions. Collaboration is about people communicating and working together on a shared goal.[5] Collaboration for sustainable innovation is a mindset, as well as a process and strategy. The mindset of collaboration towards sustainable innovation is important because the innovation process itself can be really challenging. One of the most rewarding aspects of innovation is working with different people and diverse organizations to overcome these challenges. There will invariably be ups and downs throughout the sustainable innovation process, but a mindset that's focused on working together, building partnerships and relationships based on openness and trust will be a significant advantage.

Being intentional about shaping a positive, collaborative culture that not only plays to the strengths of each person and organization but also helps everyone to achieve their goals and aspirations is a powerful enabler of a successful multidisciplinary expert team. Find meaningful, shared purpose and understand the individual objectives of everyone involved. Be open about your objectives, build trust between parties and aim to help each collaborator to get as much as what they need as possible, in addition to your focus on your own agenda.

The foundation of any high-performing team is trust and psychological safety, meaning that team members feel that they can speak up, challenge, offer ideas and play to their strengths respectfully, without fear of any negative consequences from the group or organization.[6,7] Don't be afraid of disagreement or challenge. Dealt with constructively, difference of opinion can be healthy, if it's managed well and if all parties have the attitudinal maturity to listen to each other and approach the difference with a solutions-focused mindset. If some of your partners are newer to collaboration than you are, lead by example, and be explicit about your behavioural hopes and expectations. In situations where you're unlikely to achieve consensus, use the 'disagree and commit' principle described earlier.

When focusing on working with external organizations, selecting the right collaboration partner(s) is key. Here are some key elements to consider:

1 Select a non-competitive partner. To collaborate, it's important that you and your partner(s) operate in a non-competing space or that you can each keep your competitive activities separate from your shared sustainable innovation projects.

2. Make sure that you share common goals and objectives from the sustainable innovation opportunity. Not all of those goals and objectives need to be the same or shared, but you do both need to care sufficiently about achieving the overall outcome.
3. Ensure that mutual trust, or at least the potential to build mutual trust, exists. Trust is a precursor to openness. If you trust each other, you can be open, air differences and work through any tensions and potential conflicts positively.
4. Identify complementary resources and skills. You can power up your sustainable innovation by extending your pool of resources and skills via those offered by your collaboration partner(s), and they can benefit from yours reciprocally.
5. Check leadership commitment. When you're bringing organizations together to collaborate, there needs to be commitment from the leadership of each of those organizations to make the collaboration work.
6. Ensure that there is willingness to put good processes in place. Processes need to be considered and it is essential to work through any 'what if' scenarios, sort out the 'what if' scenarios and plan for them before they happen. This will help the whole collaboration work much more smoothly.
7. The more, good-quality conversations you can have with your partner(s), right at the very outset before the collaborative innovation project starts, the better. This enables you to set the stall out and make sure that everybody's aligned. You can resolve any issues that might arise before you move forward and focus on progressing the project.
8. Collaborative sustainable innovation requires advanced leadership and teaming skills because you've got your own objectives you want to achieve and need to accommodate other parties and their ways of working, style preferences and goals. Good processes and governance help. To establish these, you'll need great communication. Plan how decisions are going to be made and by whom.

Knowledge exchange events to engage your sustainable innovation ecosystem

Knowledge exchange events can play a useful role in connecting individuals, organizations and communities to share ideas, resources and expertise. A networking event for your wider ecosystem partners can lead to valuable connections, promote collaboration and advance knowledge.

Popular formats for knowledge exchange events include panel discussions, roundtables, speed networking and mixers which help engage the wider sustainable innovation ecosystem.

Panel discussions

Panel discussions might typically feature 3–5 panellists who share their insights, experiences and opinions on the chosen subject. A moderator guides the conversation, poses questions and encourages audience participation. This format promotes an exchange of ideas, knowledge and perspectives.

To run a successful knowledge exchange panel discussion:

- Select a relevant and engaging topic that aligns with the interests of your target audience.
- Identify and invite knowledgeable panellists who can offer diverse perspectives.
- Choose an experienced moderator to facilitate the discussion and keep the conversation on track.
- Promote audience participation by reserving time for Q&A and encouraging attendees to submit questions in advance.
- Ensure the panel runs smoothly by preparing a clear agenda, setting time limits for each panellist and having a briefing session with all participants beforehand.

Roundtables

A roundtable is an event format where a small group of participants engage in an in-depth discussion on a specific topic, with everyone contributing their thoughts and ideas. Roundtables usually consist of 8–12 participants and are led by a facilitator. They are designed to encourage open dialogue, collaboration and the sharing of experiences and expertise. Participants can learn from each other, explore new ideas and build relationships in a more intimate setting.

To organize a productive roundtable event:

- Determine the topics of interest for your roundtables and select skilled facilitators.
- Arrange seating in a circular or semi-circular formation to promote a sense of equality and encourage conversation.
- Set ground rules for the discussion, such as respecting others' opinions and actively listening.
- Encourage participants to contribute their thoughts, ideas and experiences.
- Provide prompts or questions to guide the conversation and maintain focus on the topic.
- Make sure the ideas, questions and decisions generated by the participants are documented and followed up after the event.

Speed networking

Speed networking is a fast-paced event format designed to help participants make multiple new connections in a short amount of time. Similar to speed dating, participants have a series of brief, one-on-one conversations with other attendees. After a set time (usually 3–5 minutes), attendees rotate to a new partner. The following format allows for efficient and focused networking:

- Set up the event space with two rows of chairs or tables facing each other.
- Provide attendees with name tags, business cards and conversation prompts to facilitate interaction.
- Assign a time limit for each conversation and use a timer or bell to signal when participants should switch partners.

- Encourage attendees to exchange contact information and make notes about their conversations.
- Allow time at the end of the event for participants to reconnect with individuals they found particularly interesting.

Mixers

A mixer is a casual, social event designed to encourage networking and relationship building among attendees. Mixers typically involve food, drinks and a relaxed atmosphere that promotes conversation and interaction. They provide an opportunity for participants to mingle, discuss shared interests and form personal connections in an informal setting. These events can be held at various venues, such as bars, restaurants or event spaces.

To run a successful mixer:

- Choose a venue that offers a comfortable and inviting atmosphere for conversation and mingling.
- Provide refreshments, such as snacks and drinks, to create a relaxed and enjoyable environment.
- Encourage attendees to wear name tags and include their professional affiliation or area of expertise.
- Set up conversation areas or stations with topics of interest to encourage focused discussions and facilitate connections.
- Consider incorporating icebreaker activities or games to help participants feel more comfortable and engaged.
- Provide attendees with a way to exchange contact information, such as business cards or a digital platform, to develop ongoing connections after the event.

For in-person events, choose a venue that is easily accessible, has sufficient capacity and provides a comfortable atmosphere for networking. For virtual events, select a user-friendly platform that supports video conferencing, breakout rooms and chat functionality.

Methods and toolkits for multidisciplinary sustainable innovation teams: innovation and design sprints and concurrent design

Innovation sprints and design sprints provide flexible, efficient and effective methodologies and toolkits to level up and accelerate effective knowledge exchange, ideation and problem-solving within your multidisciplinary sustainable innovation team.

Innovation sprints and design thinking sprints are dynamic frameworks that empower multidisciplinary expert teams to address sustainable innovation challenges effectively. These sprints are structured, time-constrained processes that encourage teams to push the boundaries of creativity, rapidly prototype and test ideas within a span of days rather than months.

Innovation sprints begin with a clear focus on a critical challenge and move through a series of phases – understanding, ideation, decision-making, prototyping and testing. This approach allows teams to dissect complex problems, generate a wide array of potential solutions and iterate on these solutions quickly. The intense, collaborative nature of innovation sprints aligns well with multidisciplinary teams, providing a platform for each expert to contribute their unique perspective and knowledge.

Design thinking sprints, on the other hand, place a strong emphasis on human-centred design. They start with empathy, striving to understand the end user's needs and experiences. From there, teams define the problem, ideate solutions, prototype and test with users. This method ensures that the solutions developed are not only innovative but also practical and desirable from a user standpoint.

Both frameworks foster a culture of fast failure, learning and adaptation, which is critical in sustainable innovation. The sprints' iterative nature means that sustainability is considered at every stage, ensuring that the innovations not only meet current needs but do so without compromising future generations' ability to meet their own.

For multidisciplinary teams, these sprints offer structured creativity. The diverse expertise within the team is leveraged to challenge assumptions, uncover insights and drive breakthrough innovations that are sustainable both environmentally and in market viability. The sprint frameworks help teams navigate the complexities of sustainable innovation by promoting quick learning cycles and ensuring that all disciplines have a say in the solution.

Innovation sprints

An innovation sprint is a short, focused event, that gets all the right people in the physical or virtual room to collaborate on a mission.[8]

The sprint mission is a big challenge or opportunity that is significant to you, your team or your organization. It's something that you genuinely want to crack. Examples of sprint missions are:

- How might we design a digital twin to model out future sustainable energy requirements over the next five years?
- How might we deliver a truly net zero construction project?
- How might we digitally map our legacy assets for sustainable innovation planning?

Innovation sprints are designed to get the very best collective creativity and wisdom from everyone present in an efficient, dynamic and creative way. In addition to being a focused and fast way of getting results for your organization, an innovation sprint will bring the following benefits to your organization:

- Improved collaboration and teamwork.
- Creative problem-solving skills and innovation expertise.
- Better quality debate.

There are different innovation sprint processes. One of the most common is detailed in the book *Sprint* by Jake Knapp et al of Google Ventures.[9] This is a great starting point for beginners but bear in mind that the book shows just one way. It's not the only or necessarily the best way. The underlying principles of innovation sprints aren't as new as you might think. There are many similarities with the Osborn–Parnes Creative Problem Solving (CPS) Process from the

1950s. This is a structured method of coming up with creative and innovative ways to find solutions to problems and opportunities.[10]

Here is my innovation sprint step-by-step process that you can adapt and use with your team to develop and accelerate your sustainable innovation projects. It incorporates some of the elements of the processes described above.[11]

1 Bullseye session:

Fully understand and explore the innovation sprint mission, any data and different stakeholder perspectives, agree on key questions and create your sprint challenge statement.

2 Ideation:

Generate ideas towards 'solving' the challenge statement.

3 Idea selection and development:

Select the most promising ideas and improve on them.

4 Solution or prototype development (depending on what the innovation sprint is for):

Create a quick, simple sketch or 3D mock-up of what the potential solution is and how it might work.

5 Solution testing:

Use your sketch or mock-up to solicit feedback from customers, users and key stakeholders.

6 Action planning:

Agree who will do what, and by when, to progress the solution to the next stage.

At the end of your innovation sprint, you'll have several key outcomes. First, there's a prototype or sketch, a tangible but not necessarily fully functional version of your sustainable innovation, designed to test its core concepts. Alongside this, you'll have gathered valuable feedback from real users who interact with the prototype or sketch, offering insights into what works, what doesn't, and why. This feedback provides clear next steps, which might include refining the prototype, further testing or even exploring a new direction based on what you've learnt. You'll also be equipped with solid data to aid decision-making about the sustainable innovation project's future, whether that means moving forward with development, pivoting the concept or halting the project altogether. Essentially, an innovation sprint provides a fast, focused approach to tackling a challenge, experimenting with ideas and deciding on the best course of action.

Remember that subjecting the same people to the same process repeatedly means that your innovation sprints will lose their freshness and edge. Please don't stick relentlessly to a single approach.

An innovation sprint can last for anything between half a day to five days. The duration really depends on the size and complexity of the challenge that you're working on. When a sprint runs over days, it used to be that these days were back to back. Now organizations are spreading their innovation sprint activities out over two to six weeks, sometimes even longer, depending on the sprint topic.

The advantages of innovation sprinting in one hit are that it's dynamic, pacey, focused and done quickly. The benefits of spreading the sprint sessions out more are that often more people can attend, as taking a big chunk of time out of their diaries for an innovation sprint is a big ask. Work on the sprint mission can happen between sessions.

You'll certainly want to invite subject matter experts to participate in your innovation sprints. Also invite people who have limited knowledge of the technical aspects of your challenge. They will be brilliant wild cards, ask fundamental questions that will get everyone thinking differently and add value to your innovation sprint dynamic.

In addition, invite people to give lightning talks, expert interviews or fireside chats at your innovation sprint. You'll need perspectives on the challenge from key stakeholders, and these formats are great ways of introducing those activities into your sprint activities. Make some of these sessions inspirational or thought-provoking, too. You can invite people who have solved a similar or related challenge elsewhere, for example, to share their stories at your innovation sprint. These don't have to be people from inside your organization. It's even better, in fact, if some of them come from outside it. Remember that not everyone has to stay for the whole sprint. Some people can pop in, contribute and leave.

Bring trusted customers into the sprint if you can at key stages. They will give you some invaluable feedback. Hear from them about the problems and opportunities they would like you to solve at the beginning, then test your ideas with them at the end.

There's no rule around how many people should participate in an innovation sprint. You need enough people to spark and sustain creative thinking throughout the innovation sprint, mix up the groups and have some positively disruptive conversations. A minimum number for this is around 8, and you can sprint with anywhere up to 120–130 people, even more. Just be sure to be well organized.

CASE STUDY

Northumbrian Water sprinting to sustainable innovation

In 2023, the seventh Innovation Festival by Northumbrian Water marked a significant stride in collaborative innovation, drawing more than 2,500 individuals from across the UK to address 40 critical challenges affecting the water sector and broader environmental and societal issues in dedicated innovation sprints, each running simultaneously. The event's effectiveness stemmed from a collective commitment to learning, understanding and creating solutions that drive progress.

The practical outcomes of the festival ranged from incremental changes to potential breakthroughs in operational efficiency and environmental management. Contributing to the event's success was a series of presentations by inspiring speakers who provided diverse insights, helping participants to think differently. The week-long event facilitated more than 40,000 new connections, many of which are expected to evolve into collaborative projects or initiatives.

During the supply chain and ecosystem sessions, the emphasis was on leveraging partnerships with the supply chain and academic institutions to discover innovative solutions. Northumbrian Water's adoption of the open innovation model was evident, with over 650 organizations and 21 universities contributing to the problem-solving efforts.

The festival's dynamic is redefined each year, with Newcastle Racecourse providing a conducive environment for the intensive innovation work. The diversity of participants was notable, contributing to a range of creative solutions drawing from different sectors and individual experiences.

A significant number of Northumbrian Water Group employees, more than 700, participated actively in the festival. Their involvement was integral to sharing insights, identifying key issues and fostering a company culture receptive to innovation and improvement.[12]

CASE STUDY

Innovation sprints in action: enviroSPRINT 21

The enviroSPRINT 2021 event was an innovative three-day online sprint that brought together postgraduate researchers (PGRs) from across the United Kingdom, including a broad spectrum of the National Environment Research Council (NERC) Doctoral Training Partnerships (DTPs) including ARIES, EnvEast, C-CLEAR, Cambridge ESS, ENVISION, IAPETUS2, INSPIRE and SPITFIRE.

The event was a dynamic gathering of interdisciplinary talent focused on developing actionable solutions to pressing environmental challenges, signalling a significant leap towards sustainable innovation.

Throughout the duration of enviroSPRINT, interdisciplinary teams comprising PGRs focused on one of five distinct challenge areas, each embodying critical concerns for policymakers, industry and third sector organizations within the environmental sciences. The sprint was not just about ideation but involved rapid, intensive research to pinpoint key issues within these areas before moving on to the innovation phase where potential solutions were crafted.

Expertise was not in short supply. EnviroSPRINT saw contributions from various nationally significant organizations. These experts lent their knowledge and insights, offering background information and advice that enriched the teams' understanding and approach to their respective challenges. The culmination of the event was a showcase where teams presented prototypes of their solutions to a panel of expert judges. These presentations were assessed based on the novelty and feasibility of the concept, as well as the quality of presentation, with a winner being selected for their standout contribution.

The challenges tackled were diverse and critical:

- Sustainability of wetlands: Spearheaded by the ARIES and EnvEast DTPs, in collaboration with the Anglian Centre for Water Studies alongside Anglian Water and Norfolk Rivers Trust, this challenge focused on the multifaceted value of wetlands.
- Nature-based solutions to climate change: Guided by the C-CLEAR and Cambridge ESS DTPs, along with the Cambridge Conservation Initiative (CCI), this area explored the potential of nature to combat climate change.
- Greenhouse gas removal: Addressing the complex barriers to the implementation of greenhouse gas removal strategies, this challenge was led by the INSPIRE and SPITFIRE DTPs with input from industry partners.

- Soil carbon management for climate mitigation: The ENVISION DTP and the UK Centre for Ecology and Hydrology (UKCEH) tackled the critical role of soil in mitigating climate change.
- Sustainable deep-sea mining: Led by IAPETUS2 DTP, in conjunction with industry partners, this challenge sought to address the environmental implications of deep-sea mining.

The enviroSPRINT event shows the power of interdisciplinary collaboration in forging sustainable solutions. By bridging diverse fields of expertise and fostering a dynamic environment of innovation, the sprint has illuminated paths towards environmental sustainability.[13,14]

CASE STUDY

Innovation Sprints at the 2050 Maritime Innovation Hub

The 2050 Maritime Innovation Hub at the Port of Tyne, UK, has been at the forefront of integrating innovation sprints into the maritime industry with a focus on sustainable innovation. Since its inception, the Hub has aimed to address the pressing environmental challenges and technological needs of the maritime sector through a collaborative and open innovation approach.

Aligned with the UK's Maritime 2050 Strategy, the 2050 Maritime Innovation Hub has been designed to foster collaboration between industry, academia and the public sector. The Hub's dedication to sustainable innovation is rooted in the understanding that the future of the maritime industry depends on its capacity to adapt to environmental concerns, including reducing emissions, improving energy efficiency and managing the impact of maritime operations on marine ecosystems.

The innovation sprints at the Hub are intensive, focused sessions where multidisciplinary teams from different organizations in the maritime innovation ecosystem come together to brainstorm, prototype and test new ideas over a short period. Typically structured in phases, the sprints start with a deep dive into the problem space, followed by rapid ideation, the development of prototypes or solution sketches, user testing and, finally, refinement of the solutions.

Sprints at the Hub are particularly centred on sustainable innovation. Teams are tasked with developing solutions that not only drive economic growth but also ensure environmental protection and social responsibility. This might include designing new waste reduction approaches, developing clean energy solutions for ports or creating digital tools to optimize resource use and reduce the environmental footprint of maritime and supply chain operations. Innovation sprints have also helped to progress work on challenges such as implementing autonomous shipping, digital twins and the feasibility of green shipping corridors.

The potential solutions developed during the sprints serve as a foundation for further R&D, and often evolve into fully fledged projects that can be implemented within the industry. The innovation sprints at the 2050 Maritime Innovation Hub have demonstrated their effectiveness as a tool for driving sustainable innovation across the maritime sector. Through focused collaboration, knowledge exchange, rapid solution-finding and an open approach to innovation, innovation sprints at the Hub are contributing to the creation of a more sustainable and resilient maritime industry.

Design sprints

Design sprints place the emphasis on 'thinking like a designer' or 'design thinking', with the customer or end user rightly at the centre of the whole process. Both innovation and design sprints are time-bound, structured, intensive, iterative processes, often resulting in high levels of productivity. They also rely on cross-functional teams working together, incorporate user feedback and benefit from effective facilitation to guide the sprint and ensure the process stays on track.

Some differences between innovation and design sprints, however, are:

- Scope: Design sprints are customer-centric, being more focused on user experience and interface design. Innovation sprints cover a wider range of objectives.
- Outcome: The primary outcome of a design sprint is a user-tested prototype, whereas an innovation sprint may result in a broader range of deliverables, including new business strategies or operational changes.
- Flexibility: Innovation sprints can be more adaptable in their structure and duration, while design sprints typically follow a more rigid, five-day process.

- Testing: Design sprints always end with user testing, while innovation sprints may or may not include this step, depending on whether the solution is customer-facing or internal.

In practice, organizations may blend these methodologies or adapt them to suit their unique needs and challenges. The choice between a design sprint and an innovation sprint often depends on the specific problem at hand and the goal of the sprint.

The Council of Europe positions design thinking as an intentional process to generate novel, meaningful solutions that yield positive impact.[15] It's a comprehensive method for creative problem-solving, equipped with its own toolkit and emphasizing user involvement in the process.

A design sprint typically has five key stages:

- The first stage, 'empathize', necessitates a deep understanding of the customer – their thoughts, actions, feelings and objectives.
- The next stage is 'define', to identify and articulate what needs to be created for the customer, either a product or a service.
- Following this is the 'ideate' stage where ideas are generated in alignment with the defined objectives.
- Next comes 'prototype', which involves giving these ideas physical form for better understanding and evaluation.
- The final stage is 'test', where these prototypes are put to use and their performance is evaluated. The feedback received is then used to refine and possibly reiterate the entire process.

There can be considerable back-and-forth movement between stages, reflecting an iterative loop rather than a straightforward path. This flexibility is a fundamental aspect of the design thinking process, accommodating constant refinement and adaptation.

Design thinking, being so attuned to users' needs, also minimizes risk. By collaborating directly with the people who will use the product, service or solution, organizations enhance its adoption rate and increase the likelihood of its success.

CASE STUDY

Innosprint: sustainable innovation in the Estonian public sector

The Estonian Public Sector Innovation Team, known as Innotiim, embarked on a transformative journey to embed new methodologies for problem-solving within the Estonian public sector. Over an 18-month period, Innotiim honed a design sprint format, Innosprint, which became an exemplar for introducing design thinking methods to people working in the public sector.

The central challenge was upskilling public sector employees to adopt people-focused approaches and embrace design thinking methods. Innotiim faced the task of creating an experience that could train facilitators for several teams multiple times a year in a cost-effective way.

Innotiim tailored Google Venture's Design Sprint format mentioned above, and other similar methods, to the public sector's unique needs, resulting in Innosprint. This five-day event, spread over six weeks, diverged from traditional sprints by allocating time between the first and second days for participants to conduct user research. This gap allowed civil servants to step into the users' shoes, fundamentally shifting their problem-solving perspective.

Innosprint involved 381 individuals from 85 organizations forming 53 project teams, all seeking innovative solutions to real policy or service problems. The majority had no prior exposure to design thinking. Teams often included civil servants and members from the private or third sector. The sprints were centrally managed by a lean Innotiim team of four people, conducted virtually, and allowed participation from across Estonia.

The innovation directly benefited civil servants, enhancing their skill set for generating user-centric and efficient public policy. Indirectly, Estonian citizens gained from the improved quality of public services. Innosprint has been well received, with exponential growth in participation and the potential for extensive replication, given its cost-effectiveness and simplicity. The main limitation was the number of trained facilitators, a challenge Innotiim addressed with its facilitator training module.

Innosprint has led to improved procurement quality, the in-house adoption of design thinking by Estonia's public institutions and the integration of collaborative tools and co-creative methods into daily public sector work. Participants' feedback highlighted the value of applying learnt skills and expressed enthusiasm for continuing such development.

Following each Innosprint, Innotiim refined the format, with the latest addition being facilitator training. The format has matured to a replicable model, with increased interest and independent adoptions by Estonia's public institutions. Innotiim now focuses on evaluating the broader impact of Innosprint, particularly how participants have integrated sprint methodologies into their routine tasks and the subsequent success of the developed solutions.

Innotiim's credibility and the support of public sector leaders were crucial for Innosprint's success. The tangible results from each sprint inspired further participation and experimentation with the format. Continuous adaptation to meet evolving civil servant needs was key, as was the emphasis on tangible outcomes.

Innosprint stands as a robust, scalable and efficient method to elevate the public sector's problem-solving capability. It reflects a shift towards sustainable innovation, emphasizing practical learning, empowerment and the broader integration of design thinking in public service. The programme's success illustrates how structured innovation sprints can lead to significant advancements in the quality, sustainability and delivery of public services.[16]

Concurrent design

Concurrent design, also known as parallel design, can transform how multidisciplinary expert teams approach complex sustainable innovation projects. This collaborative methodology allows for the simultaneous development of different aspects of a project, breaking away from the traditional linear approach where each phase must be completed before moving on to the next. By enabling real-time feedback and iteration across disciplines, concurrent design shortens the project timeline and significantly improves the quality of the final product.

One of the most compelling advantages of concurrent design is its ability to integrate problem-solving efforts across various fields. This holistic approach is particularly vital in sustainable innovation projects, where the challenges are multifaceted, spanning environmental, social and economic dimensions. The method's efficiency and speed are invaluable for quickly addressing environmental and social issues, while its collaborative nature facilitates a creative and innovative environment. The cross-pollination of ideas among team members from different disciplines leads to more creative and effective solutions to sustainability challenges.

Implementing concurrent design in complex projects requires a blend of clear communication, the use of advanced technology, the assembly of a diverse team and the adoption of agile methodologies. Effective communication tools and protocols are needed to ensure seamless information and feedback sharing among team members. Collaborative design and project management software facilitate real-time updates, simulations and modifications. The composition of the team is critical; a broad range of expertise enriches the project development process with unique insights and skills.

CASE STUDY

Concurrent design at the European Space Agency

Concurrent design is a methodical approach that the European Space Agency (ESA) uses to streamline the design process of spacecraft and missions. The approach allows different teams to work simultaneously rather than sequentially, supporting efficiency and innovation.

Why ESA uses concurrent design

ESA adopted concurrent design to tackle complex projects more effectively. Traditional design processes, where stages occur one after the other, can be slow and prone to errors as changes in one area might necessitate revisions in another. Concurrent design addresses these challenges by enabling real-time collaboration across disciplines. This method reduces the time and cost associated with space missions while improving quality and fostering more innovative solutions.

How concurrent design works

In practice, concurrent design involves multidisciplinary expert teams working together in a dedicated space, sharing information and updates in real time. This approach relies heavily on advanced software tools that allow for the integration of various design aspects, from engineering and testing to cost analysis. These tools enable team members to immediately see the impacts of design decisions across the project, allowing for quick adjustments that optimize the mission's overall design.

ESA's special facility: the Concurrent Design Facility (CDF)

At the heart of ESA's adoption of concurrent design is its Concurrent Design Facility (CDF), located at the ESTEC site in Noordwijk, the Netherlands. The

CDF is a state-of-the-art environment designed to support the concurrent design process. It features a large central working area surrounded by smaller breakout rooms equipped with the latest in communication and design technologies.

Teams at the CDF use a standardized methodology that starts with defining the mission's requirements. From there, they work through the design phase, with specialists in various fields contributing their expertise in real time. This collaborative environment allows for the rapid iteration of designs, with immediate feedback loops that significantly shorten the design phase in comparison with traditional approaches.

Benefits and outcomes

The use of concurrent design and the facilities at the CDF have led to significant successes for ESA. Projects can move from concept to detailed design much faster than through traditional methods, reducing timescales from years to months. This efficiency not only saves time but also significantly cuts costs, as potential issues are identified and resolved much earlier in the design process.

The collaborative nature of concurrent design enhances the quality of ESA's missions. By bringing together diverse expertise and perspectives from the outset, the agency ensures that its missions are both innovative and robust. Concurrent design represents a paradigm shift in how space missions are conceived and executed. At ESA, this approach has transformed the design process. The CDF stands as a testament to ESA's commitment to embracing cutting-edge methodologies to push the boundaries of space exploration. Through concurrent design, ESA continues to lead in the development of missions that expand our understanding of the universe.[17]

Method for multidisciplinary sustainable innovation teams: charettes for stakeholder engagement

A charrette is an intensive, collaborative workshop that brings together stakeholders to rapidly create a design or vision for a project. Charrettes, due to their collaborative nature, are particularly useful in complex projects where diverse expertise and stakeholder input are crucial for creating viable, sustainable solutions. They are also effective ways of generating the community engagement that is essential for many sustainable innovation projects.

Charrettes are a platform for integrating sustainability deeply into the design and planning of an innovation project from the outset. They are like innovation sprints and design sprints in that they are time-boxed, collaborative processes aimed at solving complex problems. However, charrettes differ primarily in their focus on collective decision-making and consensus-building among a wide range of stakeholders. They are typically used in the context of community engagement for sustainable urban planning, architecture and public policy. Innovation sprints and design sprints, however, are more commonly associated with product development and business strategy. While innovation sprints often have a broader focus on developing new business models or processes, and design sprints are more narrowly focused on creating and testing a specific product or service, charrettes emphasize participatory design and community engagement, often leading to solutions that have a direct public impact.

To run a charrette, start by identifying the core problem or challenge that needs to be addressed. This often involves engaging with stakeholders to understand their needs and the constraints of the project. Once you have a clear understanding of the challenge, organize a workshop that brings together a diverse group of stakeholders, including experts from relevant fields, community members, and end users.

During the charrette, first present the sustainable innovation challenge and provide background information to ensure all participants have a shared understanding of the issues at hand. It's important to create an environment that encourages open communication and collaboration, often facilitated by a neutral party skilled in leading group discussions.

The next phase is to break participants into smaller groups to discuss the sustainable innovation challenge and brainstorm solutions. These groups would then reconvene to share their ideas with the larger assembly, seeking feedback and building upon one another's suggestions. This collaborative environment is conducive to generating a wide range of ideas and ensures that all voices are heard.

As ideas are proposed and refined, the group then begins to focus on creating more detailed designs or plans. This could involve sketching out concepts, discussing potential implementation strategies and identifying potential barriers to success. Throughout the process, it's essential to continually circle back to the sustainable innovation project's goals and constraints to ensure that proposed solutions are viable.

Towards the end of the charrette, the group works towards decisions on the best solutions or designs. This part of the process may involve voting, ranking ideas or further discussion to refine and select the most promising options.

Once a decision is made, the charrette concludes with an action plan that outlines the next steps, assigns responsibilities and sets deadlines for the implementation of the sustainable innovation. Follow-up meetings might be scheduled to ensure that the momentum from the charrette is maintained and that the project moves forward as planned.

To support sustainable innovation projects, the charrette method can be used for:

- Goal setting: A charrette can kick-start the project by helping to define clear, consensus-based sustainability goals, metrics and performance benchmarks.
- Integrated design: It enables the integration of various sustainable design aspects, such as energy efficiency, water conservation and carbon footprint reduction, into the project from the start.
- Stakeholder engagement: Charrettes involve all key players, ensuring that multiple perspectives are considered, including those of end users, technical experts and community representatives.
- Risk identification: Early collaboration in a charrette can help identify potential sustainability risks and hidden costs.

- Holistic approach: Charrettes provide a platform to explore environmental, social and governance (ESG) goals, connecting design to broader sustainability outcomes.

EXAMPLES OF CHARRETTES

Mead & Hunt sustainability charrettes

Mead & Hunt, an architecture and engineering firm, uses sustainability charrettes for goal setting and project visioning. It aligns its workshops with the AIA Framework for Design Excellence, including considerations for the LEED, WELL and Living Building Challenge certification systems.[18]

Great Salt Lake charrette

Utah State University conducted a charrette focusing on the Great Salt Lake. Teams developed ideas for water use and sustainable infrastructure design, contributing to the lake's support. The charrette process was also used to generate ideas for sustainable urban development in other Utah communities.[19]

Brookfield Sustainability Institute charrette

Over the course of several days, the Brookfield Sustainability Institute hosted hundreds of students from around the world, industry partners, community stakeholders and thought leaders at the 2023 'Brookfield Smart Sustainability Charrette'.

During 23–27 February 2023, students from KEA – Copenhagen School of Design and Technology, Politecnico di Milano, Università Commerciale L. Bocconi and the University of Toronto joined the students of George Brown College to develop smart sustainable solutions to real-world problems.

The teams tackled projects such as the Short Food Supply Chain, Climate Positive Housing Development, Wooler Area Community Organization, The Linear City Project, The Silent Community, Sustainable Tourism Barbados, The Gathering Place – A Housing Collaborative, The Chef's House Remodelling and The Future of Work Project.

Throughout these five days, the students researched, discovered and investigated smart sustainable solutions.[20]

Conclusion

An organization may have exceptional multidisciplinary experts assigned to its sustainable innovation projects, but if those experts are not skilled and supported in effective collaboration, they will not succeed in operating as a high-performing team. Leaders of sustainable innovation need to cultivate a climate of trust and psychological safety, encourage and role-model constructive challenge and not shy away from productive, creative tension. They also need to help the team to work towards better understanding of each other's disciplines, thinking styles, objectives and approaches, to find common ground and encourage a positive team dynamic.

When working with external partners, selecting the right collaborators is key. You and your collaborators need to share sufficient mutual, non-competing interests and goals and have active leadership support for the collaboration and the potential to build strong trust, good governance and effective working and decision-making processes.

Knowledge exchange, ideation and problem-solving are core competencies for sustainable innovation teams, whether working entirely within the same organization or collaborating with external partners. To support and accelerate these processes, use methods such as knowledge exchange events, innovation and design sprints, concurrent design and charrettes as structures and processes to help to navigate your sustainable innovation journey. Always keep users and key stakeholders at the front and centre of your sustainable innovation thinking.

ACTION CHECKLIST

- Review how effectively your multidisciplinary team collaborates. Specifically consider the following: trust; mutual respect and understanding of each other's disciplines, thinking preferences and approaches. Discuss these with your team and jointly agree an action plan to take these to the next level.

- Reflect on how much constructive challenge takes place within the team. Is challenge avoided, divisive or healthy and helpful? If it's not healthy and helpful, consider skills development for your team in challenging others and receiving challenge appropriately.
- Evaluate how well your organization works with current or potential external collaborators, based on mutual shared goals, drive, trust, working relationships and governance effectiveness. Take any improvement actions needed to continue, strengthen or even exit the collaboration, as appropriate.
- Consider how proactively and effectively your organization leads and participates in knowledge exchange activities within your wider sustainable innovation ecosystem. You may wish to host one or more of the knowledge exchange event suggestions in this chapter.
- Assess how well your current sustainable innovation knowledge exchange, ideation and problem-solving processes are working with your team. Introduce or refresh and improve the innovation or design sprint process to level up your team's performance.
- If your innovation or design sprint process is already finely tuned and advanced, you may wish to consider progressing to developing a concurrent design method for your team or organization.
- Consider using charrette methodology if your sustainable innovation requires high levels of external stakeholder input and engagement.

Notes

1. Southwick, K (1999) *High Noon: The inside story of Scott McNealy and the rise of Sun Microsystems*, John Wiley & Sons, Inc., p. 39
2. Baldwin, H. High technology, high pressure, *CIO magazine*, 1 December 1998
3. Amazon (2017) 2016 Letter to Shareholders, www.aboutamazon.com/news/company-news/2016-letter-to-shareholders (archived at https://perma.cc/U7D3-DPWW
4. North, J (n.d.) Knowledge exchange for research and business innovation, The Big Bang Partnership, bigbangpartnership.co.uk/knowledge-exchange/ (archived at https://perma.cc/27HE-KBSG)

5 North, J (n.d.) Collaborative innovation, The Big Bang Partnership, bigbangpartnership.co.uk/collaborative-innovation/ (archived at https://perma.cc/9G5W-XE6E)

6 Lencioni, P M (2002) *The Five Dysfunctions of a Team*, Jossey-Bass

7 Gallo, A. What is psychological safety?, *Harvard Business Review*, 15 March 2023, hbr.org/2023/02/what-is-psychological-safety (archived at https://perma.cc/Y46P-5FCW)

8 North, J (n.d.) How does an Innovation Sprint work?, The Big Bang Partnership, bigbangpartnership.co.uk/how-does-an-innovation-sprint-work/ (archived at https://perma.cc/Y8L5-V57J)

9 Knapp, J, Zeratsky and Kowitz, B (2016) *Sprint: How to solve big problems and test new ideas in just five days*, illustrated edition, Simon & Schuster

10 North, J (n.d.) Creative problem-solving process – a quick history, The Big Bang Partnership, bigbangpartnership.co.uk/creative-problem-solving-process-a-quick-history/ (archived at https://perma.cc/255W-FJJG)

11 North, J (n.d.) How does an Innovation Sprint work?, The Big Bang Partnership, bigbangpartnership.co.uk/how-does-an-innovation-sprint-work/ (archived at https://perma.cc/RLJ2-SCB2)

12 NWG Innovation Festival (n.d.) The Festival, www.innovationfestival.org/the-festival/ (archived at https://perma.cc/7RPE-MQ8J)

13 enviroSPRINT (n.d.) www.envirosprint.uk/ (archived at https://perma.cc/WW79-PYCK)

14 Author's own experience as a facilitator of the enviroSPRINT event, April 2021

15 Council of the European Union (2020) Reading References: Design Thinking, www.consilium.europa.eu/media/45594/design-thinking-sdu.pdf (archived at https://perma.cc/5LM2-ADGT)

16 Observatory of Public Sector Innovation (n.d.) Innosprint – A design sprint for the public sector, oecd-opsi.org/innovations/innosprint-a-design-sprint-for-the-public-sector/ (archived at https://perma.cc/SVP6-27QD)

17 The European Space Agency (n.d.) Concurrent design facility, www.esa.int/Enabling_Support/Space_Engineering_Technology/Concurrent_Design_Facility (archived at https://perma.cc/4DP3-E5UU)

18 Mead & Hunt (2023) Five client benefits of sustainability charrettes, meadhunt.com/sustainability-charrettes/ (archived at https://perma.cc/R4K7-9Q7K)

19 Smiel, T. Great Salt Lake charrette process highlighted at upcoming USU research landscapes, Utah State TODAY, 13 June 2023, www.usu.edu/today/story/great-salt-lake-charrette-process-highlighted-at-upcoming-usu-research-landscapes (archived at https://perma.cc/5PRN-CZY8)

20 Brookfield Sustainability Institute (n.d.) Brookfield Smart Sustainability Charrette, www.brookfieldsustainabilityinstitute.com/events/smart-sustainability-charrette (archived at https://perma.cc/L97E-67NT)

12

A new sustainable innovation project management approach for a new paradigm

CHAPTER OVERVIEW

Chapter 12 brings all the themes and activities from the previous chapters together and revisits the Sustainable Innovation Roadmap now that you have completed the journey through this book.

This chapter then presents an innovation project management approach for a new paradigm. It introduces a fresh approach to project management that caters to the specific needs and challenges of sustainable innovation in complex, technical environments. This will rarely be a linear process. Project management for future-ready programmes needs to allow for trials, iterations, learning and rapid course correction if needed. Traditional methods are too heavy, cumbersome and staid. Chapter 12 shares practical tools and methods for managing sustainable innovation projects more successfully, and for combining all the themes in this book to shape your cohesive, practical Sustainable Innovation Roadmap. It also addresses how to progress any changes you need to make to improve your organization's capability and capacity for sustainable innovation, including prioritization of actions, creating a strategy for change that takes others on the journey, influencing upwards and continuing to develop your own skills, experience and ideas as a leader of sustainable innovation.

Your Sustainable Innovation Roadmap: The journey so far

On your journey to shape your Sustainable Innovation Roadmap, you have:

- Worked on your organizational readiness for sustainable innovation, aligning your sustainable innovation goals to your organization's purpose, mission, vision and values. These goals are supported by OKRs and KPIs to help you to realize your organization's purpose, mission, vision and values. You've also identified which stage of maturity regarding sustainability that your organization is at, along with its barriers to and accelerators of sustainable innovation. In addition, you've considered using Whole Life Value (WLV) as the framework for moving towards more sustainable outcomes, assessed the effectiveness and readiness of the different levels of leadership in your organization and started to consider your strategies for building a culture of continuous improvement and success over the long term. (Chapter 2)
- Mapped potential futures for your organization, to help navigate the volatility, complexity, uncertainty and ambiguity of the world, and to better understand the context and influences in your sustainable innovation projects using horizon scanning. You've reflected on the three horizons of maintaining and defending your core business, nurturing emerging opportunities and creating entirely new possibilities, along with how your organization might attend to each of these appropriately. Your PESTEL analysis has highlighted the factors that you should consider for your sustainable innovation projects, especially in this historic time of transition, driven by climate change, decarbonization and energy priorities, as well as technological, economic, geopolitical and societal change. Stakeholder analysis and engagement have informed and tested your thinking about future priorities. Using tools such as scenario planning and transformation mapping, you've hopefully gained greater clarity on what may lie ahead and how your organization could best prepare to be ready for it. (Chapter 3)

- Developed value propositions and business models for your sustainable innovations, considering both the project and the business-as-usual (BAU) phases. You'll have adapted your value propositions for each project to meet the different priorities of each stakeholder group and thought through how you might deal with the psychology of any resistance to change that you could encounter. Your business model has circularity considerations baked in from the outset, and you've approached it from a systems thinking perspective. (Chapter 4)
- Created a sustainable innovation plan and approach for your legacy assets, including the identification of any current assets that might be getting in the way of sustainable innovation progress. This includes the preservation and protection of data. You have also begun to action appropriate quick wins to make your legacy assets operate in a more sustainable way and explored their potential for circularity. (Chapter 5)
- Shaped a plan to combine and leverage people, data and technology in your organization to drive sustainable innovation success and combine being data-driven with human insight, experience and intuition. You've assessed data skills and culture in your organization and prepared an action plan to address any areas that need improving. You've also considered how you will use data effectively and efficiently at each stage of the sustainable innovation process, potentially via cloud-based collaboration tools. You may also have investigated solutions such as big data, intelligent asset management (IAM), digital twins and building information modelling (BIM) for your own organization, depending, of course, on your industry and the nature of your projects. (Chapter 6)
- Explored how to balance sustainability, affordability, value, time and risk, and the tension between managing each of these criteria which, at times, can seem to be mutually exclusive. You've considered the strategies of satisficing versus maximizing, seen how important it is to separate your goals from the project's constraints and recognized the importance of crafting a well-constructed, clear challenge statement for each of your sustainable innovation

projects. In addition, you have a decision-making toolkit to help you and your team to work through the process of evaluating different potential solutions. (Chapter 7)

- Scrutinized your organization's approach to corporate and sustainable innovation project governance and identified any potential areas for improvement. This includes risk management, further developing your OKRs and KPIs and shaping your balanced scorecard, making sure that you have the right levels of delegated authorities and independent assurance in place, are carrying out effective risk management and are acting as a capable, collaborative client with you key suppliers. (Chapter 8)
- Created your sustainable innovation strategy, having made decisions around your key strategic choices and ensuring a healthy ratio of core, adjacent and transformational projects. In so doing, you've weighed up how each strategic choice relates to the others, built a prioritized sustainable innovation pipeline and decided how you will track and manage progress. You've also identified potential barriers to and enablers of your internal innovation processes, along with strategies to deal with them. (Chapter 9)
- Carried out a strategic review of your sustainable innovation ecosystem, focusing on co-creating the conditions of mutually beneficial success both within and external to your organization. You've reviewed the strength of your innovation team and organizational culture and identified actions to improve both, as well as thinking about how you might better leverage sustainable innovation from your supply chain. Plus, you've developed a target sustainable innovation ecosystem map to work towards. (Chapter 10)
- Put in place approaches, methodologies and toolkits to support knowledge exchange, ideation and problem-solving across your internal and mixed external multidisciplinary expert teams. This includes shaping a team dynamic of trust, psychological safety, mutual respect, communication and constructive challenge. You've assessed your current and potential external partners against important criteria for positive, effective partnership working and identified potential opportunities to fruitfully employ methods

such as knowledge exchange events, innovation and design sprints, concurrent design and charrettes into your activities. (Chapter 11)

The final stage along your Sustainable Innovation Roadmap journey is to make sure that you have productive, helpful and appropriate project management processes in place to help you to implement your sustainable innovation goals to time, budget and quality.

Rethinking project management for sustainable innovation

Sustainable innovation involves creative thinking aimed at uncovering new opportunities and tackling existing problems with fresh perspectives. Sustainable innovation is a continuous, value-driven process. Project management today is evolving to meet the needs of a hybrid work environment, the period of historic transition that we are in and to focus even more on collaboration and adaptability to respond to the challenges that these changes bring.

Project management has traditionally been viewed as primarily focused on planning, organizing and delivering projects on time, within budget and to the required standard. While these are crucial functions, they only represent some aspects of project management. Owing to, not despite, the world's future being volatile, uncertain, complex and ambiguous (VUCA), project management continues to be essential. However, we need to think about how well our current project management approaches are working for us.

Integrated, collaborative and flexible project management frameworks

Those organizations who are successfully driving sustainable innovation today moved away some time ago from traditional, siloed approaches to project management. They progressed to more integrated, flexible and collaborative frameworks to deliver impact. The future of successful project delivery, particularly for sustainable innovation projects in a

VUCA world, lies in building strong partnerships, facilitating open communication and aligning objectives across all project stakeholders. By doing so, projects are more likely to achieve their goals, overcome challenges and deliver sustainable value in an increasingly complex and uncertain global and local landscape.

EXAMPLES OF INTEGRATED, COLLABORATIVE AND FLEXIBLE FRAMEWORKS IN INFRASTRUCTURE

Project 13, Integrated Project Delivery (IPD) and Alliancing are example approaches to the evolving landscape of project management, especially in the context of delivering sustainable innovation projects in a VUCA (volatile, uncertain, complex and ambiguous) environment. These frameworks and methodologies indicate a shift towards more collaborative, transparent and flexible project management approaches. They underline the importance of strategic partnerships and shared objectives among all stakeholders.

Project 13 – UK

Project 13 is an innovative initiative developed by the UK's Institution of Civil Engineers (ICE), focused on transforming the delivery and outcome of the sector's infrastructure projects. It aims to improve outcomes for infrastructure owners, the supply chain and ultimately for the end users and society by creating a more efficient, sustainable and effective delivery model for infrastructure projects. Project 13 demonstrates a shift from traditional transaction-based approaches to a more collaborative, effective and sustainable model of project delivery.

Unlike conventional project delivery methods that operate within siloed teams, Project 13 advocates for an enterprise model. This model advocates an integrated approach where clients, contractors, suppliers and other stakeholders work together as a single, cohesive unit to enhance efficiency, innovation and value throughout the lifecycle of infrastructure projects.

Project 13 also prioritizes long-term value and performance over the lowest cost at the point of delivery. This approach aims to ensure that infrastructure investments deliver maximum benefit to society, the economy and the environment. It also encourages the use of digital tools and systems to improve the planning, design, construction and maintenance of infrastructure.

Project 13 includes a focus on developing the skills and capabilities of the workforce involved in infrastructure projects to shape a culture of continuous learning and improvement, increasing the industry's capacity to deliver complex projects successfully. It also aims to establish long-term relationships among stakeholders based on trust, shared objectives and mutual benefits. Risks and rewards are shared. This collaborative environment is designed to encourage sustainable innovation and allow for more flexible and adaptive project management.[1]

Integrated Project Delivery – USA

Integrated Project Delivery (IPD) is primarily used in the United States, where it was developed as a collaborative project delivery method to improve the efficiency and outcomes of construction projects. The approach aims to optimize project results, increase value to the owner, reduce waste and maximize efficiency through all phases of design, fabrication and construction.

Like Project 13, IPD is characterized by a shared risk and reward model, mutual respect and trust among all team members, collaborative innovation and decision-making, and open communication. The fundamental idea behind IPD is that, when key project stakeholders collaborate from the project's inception, they can more effectively identify and address issues, leading to a more efficient and streamlined project delivery process.

Unlike traditional project delivery methods where roles are rigidly defined and responsibilities are siloed, IPD encourages an integrated approach where the owner, architect, general contractor, engineers and key subcontractors work as a unified team throughout the life of the project. This unity allows for a more holistic approach to project design and construction, ensuring that the project is aligned with the owner's goals, completed on time and within budget.

IPD agreements often involve a single multiparty contract that defines the roles, responsibilities and expectations of all parties involved. This contractual arrangement is designed to support a cooperative environment that promotes cost transparency, efficiency and sustainable innovation.

By aligning the interests of the project team with the interests of the owner, IPD helps to mitigate conflict, reduce changes and claims, and deliver a higher quality product. It demonstrates a paradigm shift in project delivery, emphasizing collaboration and integration over competition and segmentation. This approach not only enhances the project outcome but also contributes to a more satisfying experience for all stakeholders involved.[2]

Alliancing – Australia and New Zealand

Alliancing, like Project 13 and IPD, involves a collaborative, non-adversarial approach to project management and delivery. It requires the project owner, contractor and key suppliers to work together in an alliance to deliver projects. The focus is on transparency, shared risks and rewards, and achieving best-for-project outcomes. This approach is particularly effective in managing the uncertainties and risks inherent in complex and innovative projects by fostering a culture of joint problem-solving, innovation and shared responsibility.

The success of Alliancing depends on the selection of partners who share compatible values and objectives, and on the establishment of robust governance structures to guide the project's strategic direction. Effective leadership and a commitment to collaborative practices are also essential to harness the full benefits of Alliancing. As such, Alliancing is a shift away from transactional relationships towards a more integrated, cooperative model of project delivery, providing a route to achieving higher quality outcomes, sustainable innovation and more efficient project delivery.[3]

Whether your sustainable innovation projects are delivered by a wholly internal team or involve working with external suppliers and partners, the essential principles for successful project management are collaboration, integration, flexibility and shared risk and reward.

Project management methodology

The choice of project management methodology can significantly influence the success or failure of a sustainable innovation project. Each methodology comes with its own set of principles, practices and suitability for handling the dynamism and unpredictability inherent in innovation.

There are five distinct project management methodologies that are used today for managing complex technical sustainable innovation projects: waterfall, agile scrum, agile kanban, wagile and obeya.

Waterfall

The waterfall methodology, one of the earliest approaches to project management, is characterized by its linear and sequential design. It divides the project into distinct phases, where each phase must be completed before the next one begins and there is little to no overlap between phases. The approach is predicated on the assumption that project requirements can be fully defined at the outset and will remain largely unchanged throughout the project lifecycle.

While the waterfall methodology offers clear structure and documentation, its rigidity makes it less suited to projects in a VUCA world. Sustainable innovation projects, by nature, come with evolving requirements and the need for adaptability as projects progress and new information becomes available. The waterfall methodology's lack of flexibility to accommodate changes without significant time and cost implications renders it generally unsuitable for many of today's complex technical sustainable innovation projects. However, many organizations continue to use it because they have not updated their processes and methods to better suit current and future needs.

Agile scrum

Agile methodologies include a broad range of frameworks and practices that adhere to the values and principles outlined in the *Agile Manifesto*, which originated in software development and emphasizes flexibility, collaboration, customer satisfaction and the ability to adapt to change.[4] Agile scrum is one of the most widely recognized and implemented Agile methodologies and is now used in a wide range of sectors and industries beyond software development. It is designed to address many of the limitations of the waterfall model.

Scrum breaks down projects into short, manageable iterations known as sprints (not to be confused with the innovation and design sprints discussed in the previous chapter), typically lasting two to four weeks, allowing teams to adapt and make changes more easily. This iterative approach, combined with regular reviews and retrospectives, makes scrum well suited to projects where requirements are expected to change or evolve. The inclusion of roles such as the product owner and scrum master ensure that the project remains aligned with user needs, and that the team is supported in its agile practices. Agile scrum is particularly effective for complex technical sustainable innovation projects in a VUCA environment, as it allows for rapid pivoting in response to new insights, technological advancements or shifts in stakeholder needs.

Agile kanban

Agile kanban is another agile methodology that focuses on visualizing work, setting a maximum number of tasks, which can be in any one stage of the workflow at a time to avoid overloading the team, and maximizing efficiency (or flow). Kanban boards are used to visualize the flow of tasks, allowing teams to see the status of work at any given time and adjust as needed. Unlike scrum, kanban does not work in fixed iterations, allowing for a more continuous flow of work. This continuous delivery model can be advantageous for sustainable innovation projects that require ongoing development and flexibility. The ability to adjust priorities on the fly and the focus on just-in-time delivery of value make agile kanban suitable for managing projects in environments characterized by uncertainty and rapid change.

Wagile

Wagile, a hybrid approach that combines elements of waterfall and agile methodologies, aims to leverage the structured planning and predictability of waterfall with the flexibility and adaptability of agile. This approach is often adopted in situations where stakeholders require the rigour and documentation of waterfall, but the project still benefits from the iterative testing and feedback loops of agile.

Wagile can be particularly useful for complex technical sustainable innovation projects that have certain non-negotiable requirements or regulatory milestones that must be met within specific timelines but also need the adaptability to respond to evolving project dynamics. However, the success of the wagile approach depends on the team's ability to balance the structured aspects of waterfall with the iterative nature of agile, which can be challenging to implement effectively.

Obeya

Obeya, which translates to 'large room' in Japanese, is a lean project management approach that emphasizes transparency, cross-functional collaboration and visual management. Teams create an obeya room which serves as a physical or virtual space where all project information, such as plans, metrics and progress, is displayed visually for team members to see. This approach facilitates a holistic understanding of the project, encourages real-time communication and supports decision-making. For complex technical sustainable innovation projects, obeya can be highly effective in aligning team members, breaking down silos and ensuring that everyone is focused on the collective goals of the project. The visual nature of obeya and its emphasis on collaboration make it well suited to managing projects in a VUCA world, where clarity, agility and team cohesion are essential.

The optimal project management methodology for your organization

While no single project management methodology is universally suitable for all types of sustainable innovation projects, agile scrum and agile kanban stand out for their flexibility, adaptability and emphasis on rapid iteration and feedback. Wagile and obeya offer valuable hybrid and visual approaches, respectively, that can be tailored to specific project needs, blending structured planning with agility and collaboration. The traditional waterfall methodology, with its linear and rigid structure, is generally less suited to the dynamic and uncertain nature of sustainable innovation projects, where requirements and external conditions can change rapidly. Ultimately, your choice of methodology should be guided by the project's specific characteristics, including its goals, complexity, stakeholder needs and the degree of uncertainty involved.

It's also important to remember that you can tailor your selected methodology or combine it with elements of other methodologies to best suit your organization's requirements and the team's working style. Just make sure that everyone involved is clear about the project management method and any adaptations, as well as their own and each other's roles in the process. Make sure that your process is documented, supported by the relevant templates and collaborative tools, followed and reviewed regularly to ensure continued fitness for purpose.

PROJECT MANAGEMENT METHODOLOGY CHECKLIST

This checklist is designed to guide you through a series of questions to help determine which project management methodology – waterfall, agile scrum, agile kanban, wagile or obeya – is most suitable for your project's unique needs.

Project complexity and scope

- Are the project's goals and requirements well defined and unlikely to change?

(Waterfall might be suitable)

- Does the project involve high levels of complexity and uncertainty, requiring frequent reassessment and adaptation?

(Agile scrum or agile kanban might be suitable)

- Is there a need for continuous delivery and flexibility in task prioritization without fixed iterations?

(Agile kanban might be suitable)

Team structure and dynamics

- Does the project team favour structured phases with clear milestones and deliverables?

(Waterfall or wagile might be suitable)

- Is the team experienced in agile methodologies and capable of self-organization?

(Agile scrum might be suitable)

- Does the project benefit from visual management and real-time collaboration among cross-functional teams?

(Obeya might be suitable)

Stakeholder engagement and feedback

- Are stakeholders expecting regular updates and opportunities to provide feedback?

(Agile scrum or agile kanban might be suitable)

- Is there a requirement for high stakeholder involvement throughout the project lifecycle?

(Agile scrum or obeya might be suitable)

Risk management and flexibility

- Does the project need to accommodate high risk and frequent changes in scope or priorities?

(Agile scrum or agile kanban might be suitable)

- Is there a blend of fixed requirements and the need for adaptability to new information or technology?

(Wagile might be suitable)

Delivery timeline and budget

- Is the project timeline fixed with a well-defined budget?

(Waterfall or wagile might be suitable)

- Does the project require flexibility in delivery timelines to accommodate evolving features or solutions?

(Agile scrum or agile kanban might be suitable)

Cultural fit and change management

- Does the organizational culture support a collaborative, transparent approach to project management?

(Agile scrum, agile kanban or obeya might be suitable)

- Is the organization or project team new to agile methodologies, requiring a more gradual transition?

(Wagile might be suitable)

Note: as stated above, remember that you can tailor your selected methodology or combine it with elements of other methodologies to best suit your organization's requirements and the team's working style.

Your change-making strategy

Depending on the specific circumstances of your own organization, the work that you've been doing as you've progressed through the chapters may have left you with minor tweaks and enhancements or some significant changes to make in a few areas. You might even need a complete overhaul of your current setup to deliver your sustainable innovation projects. If it's either of the latter two, you might feel excited by the challenge or overwhelmed. It can be tempting to try to change everything at once, but a prioritized approach focusing on quick wins and starting work on only the most important changes will generally lead to less confusion, be more manageable and gain greater traction in your organization. People in busy roles can only accommodate so much change at once, so pick your most beneficial and important improvements and go with those first. Budgets permitting, you may also want to consider bringing in additional, temporary resource to support you through the process and create capacity to build new approaches at the same time as delivering the day job and responding to emerging, immediate opportunities.

People respond better to change that they have been involved in creating than they do to change that has been 'done to' them. Endeavour to take others with you on the journey, including using methods such as the innovation and design sprints presented in the previous chapter.

Influencing upwards

One of your challenges may be influencing upwards and navigating your organizational hierarchies to persuade senior executives and board members of the strategic importance of any changes that you've identified. You'll need to use a strategic blend of communication, evidence-based advocacy, and the demonstration of alignment with the organization's broader strategic goals.

Communication

Of course, to influence upwards you'll need to articulate a compelling vision that encapsulates the value of improving how your organization is set up to deliver sustainable innovation, not just from an environmental and social perspective but also in terms of long-term business viability and competitive advantage. This involves translating sustainable innovation process goals into the language of business outcomes – such as risk mitigation, market differentiation, customer loyalty and growth opportunities. By presenting your proposed sustainable innovation improvements as an essential part of the organization's strategic objectives, you're more likely to gain executive support and commitment because others will be better able to understand how they are also intrinsically linked to organizational success.

Evidence-based advocacy

Present concrete data and case studies that demonstrate the positive impact of improving the organization's capability and capacity to deliver sustainable innovation. This might include success stories from within the organization or benchmarks from industry peers who have successfully implemented the improvements you're proposing and seen tangible benefits. By showcasing evidence of reduced costs, enhanced brand reputation, increased operational efficiency or new market opportunities, make a compelling case for the changes you recommend.

Alignment with organizational goals

Illustrate how the changes you're proposing could also align with and support other strategic objectives of the organization. For example, could the changes also benefit work on other priorities, such as growth, customer satisfaction, regulatory compliance or digital transformation?

Championing process and governance changes

Influencing upwards to advocate for necessary process and governance changes involves identifying and addressing potential bottlenecks or inefficiencies in the current system that hinder sustainable innovation. Propose specific, actionable changes that could facilitate smoother project execution, and why, rather than generalizing about the changes that are needed. Present a clear plan for how these changes can be implemented, and how they will lead to improved sustainable innovation project outcomes.

Advocating for team and organizational development

You may also need to champion the development of skills and competencies required for sustainable innovation within your team and the wider organization. This might involve advocating for training programmes, new hiring practices that prioritize sustainable innovation expertise, or the creation of dedicated innovation roles or departments. Put forward a clear business case, complete with financial and non-financial costs and benefits, and an outline of an implementation plan.

Building coalitions and demonstrating success

Build coalitions of supporters across the organization who can collectively advocate for the sustainable innovation improvements and changes that need to be made. Seek allies among other leaders, influential stakeholders and even external partners who share your vision.

Demonstrating quick wins and early successes from some smaller quick wins can also help build momentum and help you to influence upwards.

Your own leadership development

The most successful leaders of sustainable innovation constantly seek ways to refine their strategies and approaches to stay ahead. In addition to always working on creating and communicating a strategic vision, and building dynamic capability in your organization, actively participate in and contribute to forums, think tanks and consortiums that are at the forefront of sustainable innovation.

By engaging with these ecosystems, you can gain early insights into emerging technologies, regulatory changes and shifting market dynamics. You'll be more able to anticipate trends rather than react to them and position your organization to leverage opportunities and mitigate the risks associated with sustainable innovation.

Role-model a culture of intellectual curiosity within your organization. Encourage team members to explore interdisciplinary knowledge and pursue learning opportunities outside their immediate technical or functional areas to help them consider more creative solutions to complex problems. You can support this by facilitating access to diverse learning resources, organizing cross-disciplinary workshops and rewarding innovative thinking that combines insights from different fields.

Another subtle yet powerful practice is the strategic use of storytelling to articulate your vision and the value of your sustainable innovation initiatives, and the changes you think the organization needs to make to deliver them. Stories that highlight the tangible impact of these initiatives and changes on the environment, society and the organization itself can resonate deeply with employees, stakeholders and customers. Inspire commitment and action across the organization by creating narratives that connect the technical aspects of sustainable innovation with its broader implications. Work with your communications team to extend this storytelling beyond internal communications so that your organization's sustainable innovation efforts are presented to external audiences, enhancing its reputation and influence in the market.

Embrace and role model a philosophy of radical collaboration yourself. In fact, you'll want to role model all the behaviours, values and approaches that you're asking of others. This goes beyond your direct team to include how you co-create with customers, suppliers, academia and even competitors. Radical collaboration requires a shift in mindset from viewing other entities as competitors or external parties to seeing them as potential partners in achieving a common goal. This can accelerate the development and implementation of sustainable innovations, creating value not just for the organization but for communities and the planet, too.

Stay prepared to lead sustainable innovation by looking beyond conventional leadership practices. Embed yourself in relevant ecosystems, develop intellectual curiosity in yourself and others, use storytelling and embrace radical collaboration to drive meaningful and lasting positive change.

Conclusion

It's important that you reassess the fitness for purpose of your current sustainable innovation project processes and identify any changes that will make your process more effective and efficient in the face of increased change and uncertainty. After working through all the chapters in this book, you may find that there are several or many things that you want to introduce or change to update your organization's approach to achieving sustainable innovation. It's important to be neither overwhelmed nor overly ambitious about trying to achieve too much at once. Make a prioritized action plan focused on quick wins. Start to progress longer term items now that will bring significant benefits later. Consider your strategy for change, being sure to take people with you on the journey. Also be strategic about how you will influence upwards, sharing a compelling vision and rationale for change. Finally, always do your best to role-model and reward the values, behaviours and approaches that you want to see in others. This includes demonstrating a commitment to continuous learning, collaboration and staying curious. Don't forget to download all the resources that go with this book: www.koganpage.com/LSI. Thanks for reading, and all the best with your sustainable innovation endeavours.

ACTION CHECKLIST

- Make sure that you have downloaded all the resources that accompany this book.
- Review all your work from the previous chapters and update with any further thoughts, questions and ideas.
- Evaluate the fitness for purpose of your current project management methodology. Identify and document any changes and improvements you want to make, and why.
- Step back from all your work so far and identify your priorities for change, including quick wins and longer-term items that you need to start working on now. Create an outline plan for how you might achieve these changes.
- Consider your change-making strategy and create a plan to take people on the journey with you. Include any upwards influencing that may be needed.
- Create a personal development plan to continue to grow your skills and external connection as a leader. Keep your personal development plan updated to always stay relevant and current.

Notes

1 Project 13 Network (n.d.) Project 13, www.project13.info/ (archived at https://perma.cc/Y2Q7-XQMM)

2 American Institute of Architects (2007) Integrated Project Delivery: A Guide, info.aia.org/siteobjects/files/ipd_guide_2007.pdf (archived at https://perma.cc/RHU2-8XTN)

3 AECOM (n.d.) Alliancing – A Participants Guide, aecom.com/alliancing/ (archived at https://perma.cc/7RXD-M6LX)

4 Manifesto for Agile Software Development (2001) agilemanifesto.org/ (archived at https://perma.cc/GFD9-DGU4)

INDEX

NB: page numbers in *italic* indicate figures or tables

Looking for another book?

Explore our award-winning books from global business experts in Responsible Business

Scan the code to browse

www.koganpage.com/responsible-business